29.95

Be Bold

Chaotic Dynamics
and Fractals

Chaotic Dynamics
and Fractals

Edited by

Michael F. Barnsley

School of Mathematics
Georgia Institute of Technology
Atlanta, Georgia

Stephen G. Demko

School of Mathematics
Georgia Institute of Technology
Atlanta, Georgia

1986

ACADEMIC PRESS, INC.
Harcourt Brace Jovanovich, Publishers
Orlando   San Diego   New York   Austin
London   Montreal   Sydney   Tokyo   Toronto

Academic Press Rapid Manuscript Reproduction

This is Volume 2 in
NOTES AND REPORTS IN MATHEMATICS IN SCIENCE AND
ENGINEERING
Edited by WILLIAM F. AMES, *Georgia Institute of Technology*

A list of books in this series is available from the publisher on request.

ACADEMIC PRESS, INC.
Orlando, Florida 32887

United Kingdom Edition published by
ACADEMIC PRESS INC. (LONDON) LTD.
24–28 Oval Road, London NW1 7DX

LIBRARY OF CONGRESS CATALOG CARD NUMBER: 85-48295

ISBN 0–12–079060–2

PRINTED IN THE UNITED STATES OF AMERICA

87 88 89   9 8 7 6 5 4 3 2

Contents

III. Applications

Contributors

Numbers in parentheses indicate the pages on which the authors' contributions begin.

Marjorie A. Asmussen (243), *Department of Mathematics, and Department of Genetics, University of Georgia, Athens, Georgia 30602*

Michael F. Barnsley (53), *School of Mathematics, Georgia Institute of Technology, Atlanta, Georgia 30332*

D. Bessis (69), *School of Mathematics, Georgia Institute of Technology, Atlanta, Georgia 30332*

Paul Blanchard (181), *Department of Mathematics, Boston University, Boston, Massachusetts 02215*

Louis Block (113), *Department of Mathematics, University of Florida, Gainesville, Florida 32601*

Bodil Branner (169), *Mathematical Institute, The Technical University of Denmark, DK-2800 Lyngby, Denmark*

N. Chafee (69), *School of Mathematics, Georgia Institute of Technology, Atlanta, Georgia 30332*

Bernard Derrida (229), *Service de Physique Théorique, CEN-Saclay, 91191 Gif-Sur-Yvette Cedex, France*

Robert L. Devaney (141), *Department of Mathematics, Boston University, Boston, Massachusetts 02215*

A. Douady (155), *48 rue Monsieur le Prince, 74006 Paris, France*

Joseph Ford (1), *School of Physics, Georgia Institute of Technology, Atlanta, Georgia 30332*

John H. Hubbard (101), *Department of Mathematics, Cornell University, Ithaca, New York 14850*

John Mallet-Paret (263), *Division of Applied Mathematics, Brown University, Providence, Rhode Island 02912*

Pierre Moussa (215), *Service de Physique Théorique, Centre d'Etudes Nucléaires de Saclay, F-91191 Gif-sur Yvette Cedex, France*

Roger D. Nussbaum (263), *Department of Mathematics, Rutgers University, New Brunswick, New Jersey 08903*

Lawrence Turyn (287), *Department of Mathematics and Statistics, Wright State University, Dayton, Ohio 45431*

Stephen J. Willson (123), *Department of Mathematics, Iowa State University, Ames, Iowa 50010*

W. D. Withers (203), *Department of Mathematics, U.S. Naval Academy, Annapolis, Maryland 21402*

Preface

This collection of papers deals with the research area of chaos, dynamical systems, and fractal geometry. Here, current interest by the scientific and mathematical community is intense, and promises to increase over the next few years, alongside the growing usage of computer graphical techniques. This volume will be required by mathematicians, physicists, and other scientists working in, or introducing themselves to, the subject. As reference material, it is of particular usefulness because the information is placed at the present frontier of research and because, due to the quality of the contributors, a significant number of the papers will be of long-standing importance.

These papers are the proceedings of a conference on chaotic dynamics held at the Georgia Institute of Technology on March 25–29, 1985. They are grouped into three collections: chaos and fractals, Julia sets, and applications. The paper by Ford introduces the chaos and fractals collection by discussing the nature of chaos. He asks what does deterministic mean? Is it enough to state that a system possesses a unique trajectory, given the constraints of finite precision and algorithmic complexity? Alas, he says, algorithmic pathology occurs as frequently as does chaos, and chaos is as common as daffodils in spring. The paper by Barnsley gives a geometric tool for succinctly describing some strange attractors and other complicated sets of data associated with chaotic systems. Bessis and Chafee consider the Henon–Hiles Hamiltonian with complex time. They argue that singularities are distributed on many sheets of the associated Riemann surface, such that their projection back in the plane forms a fractal. John Hubbard also considers a Henon family of maps from $\mathbb{C}^2$ into itself and shows that one can bring to bear some ideas for treating analytic polynomial maps from $\mathbb{C}$ into itself. This is a very important paper because it connects Julia set theory to a physical dynamical system. Louis Block introduces the idea of turbulent maps in the course of presenting new results on iteration of continuous maps from the unit interval to itself. S. Willson shows a way in which some fractals can be generated via cellular automata.

The second group of papers deals principally with complex analytic dynamics and associated fractal geometry. Robert Devaney illustrates bursts into chaos, whereby nowhere dense Julia sets suddenly become the whole complex plane as a parameter is varied, ensuring sudden wild behavior of the dynamical system everywhere. Adrien Douady gives us remarkable new algorithms for obtaining geometrical and combinatorial information about the Mandelbrot set for $z^2 - \lambda$. A key reference which he does not give is A. Douady and J. Hubbard, *Comptes Rendus* (Paris) 294 (1982) 123–126. Bodil Branner (and John Hubbard) describes the parameter space for iterated cubic polynomials. Closely related is the work of Paul Blanchard on the symbolic dynamics of iterated cubics when the Julia set is disconnected. Els Withers shows how Julia sets can be differentiated with respect to a parameter in the associated rational map, thus permitting the formulation of Taylor series expansions for the sets. Pierre Moussa describes beautiful arithmetic properties of Julia sets for polynomials with integer coefficients.

The final group of papers we call applications. Bernard Derrida shows that the partition function for various Hierarchical Lattice Ising Models obeys a functional equation which implies that its zeros accumulate on a Julia set. This yields implications and approximations for thermodynamical properties of the system, especially near critical points where phase transitions occur. Marjorie Asmussen presents models for natural selection which involve the dynamics of components of genetic material. She is precise, and argues that chaotic cycling may occur and be important. This is a fascinating area of application for dynamical systems theory. John Mallet-Paret and Roger Nussbaum examine the relation between the delay equation

$$\epsilon \dot{x}(t) = -x(t) + f(x(t-1)) \tag{1}$$

and the iterated map defined by

$$x - f(x) = 0. \tag{2}$$

They show that the dynamical structures of (2) are not presented in (1) even for small values of ϵ. Lawrence Turyn considers the forced Fisher's equation $\dot{u} = u_{xx} + f(u) + g(x - ct)$, and gives some conditions under which chaotic solutions occur.

In addition to the lectures at the conference, represented by the papers in this volume, there was a series of eight lectures by James A. Yorke which were both introductory and deep, and comprehensible to the mixed audience of mathematicians, physicists, and other scientists. Topics included statistical uncertainty in deterministic systems due to fractal basin boundaries, the structure of attractors, collisions between attractors and basin boundaries, Liapunov numbers, quasiperiodic systems and the Ruelle–Takens–Newhouse theorem, numerical experiments on dynamical systems, Smale horseshoes, and cascades of bifurcations. Illustrations were made with differential equations for damped driven pendula, the kicked double rotor, a forced electrical circuit (Van der Pol equation), the Lorenz equations, the Henon maps, Robert Shaw's dripping faucet experiment, and the Smale cat map.

Thanks go to Bill Ames, the Director of the School of Mathematics, for providing the initial funds for the conference and to the USARO, NSF, and ONR for their generous financial support under grants DAAG29-85-M-0039, DMS-8419995, and N00014-85-G-0119, respectively. Special thanks are due to Annette Rohrs for typing the complete manuscript in her usual superb way.

CHAOS: SOLVING THE UNSOLVABLE, PREDICTING

THE UNPREDICTABLE!

Joseph Ford

School of Physics
Georgia Institute of Technology
Atlanta, Georgia

I know that most men, including those at ease
with problems of the greatest complexity, can
seldom accept even the simplest and most
obvious truth if it be such as would oblige
them to admit the falsity of conclusions they
reached perhaps with great difficulty, conclusions which they have delighted in
explaining to colleagues, which they have
proudly taught to others, and which they have
woven, thread by thread, into the fabric of
their lives.

Leo Tolstoy

PRELUDE

Notions of determinism, existence-uniqueness, and exact

analytic solutions have dominated scientific thought for cen-

turies. These notions are so pervasive and so widely accepted

that, even when no more than the mere existence-uniqueness of

a solution is known, scientists feel every confidence in the

meaningfulness of symbolically writing the exact solution as

$S(t) = F(S_0, t)$, where S is the exact system state at time t

which evolved from the precise initial state S_0 according to

an unstated functional rule F implicitly defined through

existence-uniqueness. The theoretical beliefs which support

1

this formalism are primarily two in number. Physicists assume
that they can, in principle, compute and/or measure the variables S, S_0, and t to arbitrarily high precision, and mathematicians assume that they can, in principle, always construct
that which exists. Chaos theory seriously erodes these notions,
but contemporary nonlinear dynamics can lay no claim to first
discovery. Listen to James Clerk Maxwell: "If therefore...the
cultivators of physical science...are led...to the study of
singularities and instabilities, rather than the continuities
and stabilities of things, the promotion of natural knowledge
may tend to remove that prejudice in favor of determinism
which seems to arise from assuming that the physical science
of the future is a mere magnified image of that of the past."
Listen to Max Born: "But is it certain that classical
mechanics in fact permits prediction in all circumstances...?
If we require determinacy for all times,...has this any physical meaning? I am convinced that it has not, and that
Newtonian systems...are in fact indeterminate."

On the mathematical side, early in this century L. E. J.
Brouwer railed mightily against the use of nonconstructive
proofs to establish, among other things, existence and uniqueness theorems. Specifically, he rejected the law of the
excluded middle as well as proof by contradiction, proposing
that a category between true and false exists. Unfortunately,
Brouwer could never exhibit his proposed category in the
middle, and his movement languished. However, an independent
discovery by Kurt Godel may provide a third category. If an
existence-uniqueness theorem is proved by contradiction, then
the theorem is most certainly not false, but the truth of the

theorem can still come in two flavors. The theorem is true
and can be constructively proved, or the theorem is true but
is unprovable by constructive means. Let us seek to illuminate
the significance of this somewhat subtle distinction.

Consider the existence theorem, "Question Q has an answer."
Proof by contradiction insures only that an answer exists.
Subsequent proof by construction supplies the answer. Alternatively, if the subsequent proof shows that the answer cannot
be constructed, then the notion of existence becomes as
diaphanous as cotton candy, for, while the answer may exist,
humans can never know it. How does such a situation arise?
Even though question Q be quite simple, the answer may nonetheless be so complicated that it contains more information
than does the entire logical system being used to seek the
answer. But, as Gregory Chaitin has proved, a one pound theory
can no more produce a ten pound prediction than a one-hundred
pound woman can birth a two-hundred pound child. Such is the
meaning of Godel's theorem in physics.

In the past we have, without hesitation, regarded a problem as well-defined whenever it lay in the province of some
existence-uniqueness theorem. But given the preceding paragraphs, is existence-uniqueness in fact enough? Specifically,
using slightly more precise language than before, suppose we
may show that no finite algorithm, no finite functional rule
exists to compute solutions for a problem well-posed in terms
of existence-uniqueness; is this problem still well-defined?
Perhaps not; but surely, a reader may suggest, such algorithmic
pathology must be quite exceptional. Alas, it occurs as frequently as does chaos, and chaos is as common as daffodils in

spring. We can therefore no longer ignore the problem of
solving the unsolvable, predicting the unpredictable. But
before addressing this issue, let us consider a simple but
celebrated problem whose easily obtained explicit solution is
not computable by any finite algorithm. This example also
illustrates the occurrence and meaning of chaos.

1. CHAOS: AN ILLUSTRATIVE EXAMPLE

Few equations have received more attention in recent years
than the discrete logistic equation

$$Y_{n+1} = \lambda Y_n (1 - Y_n) \tag{1}$$

which maps the unit interval upon itself when $0 \le \lambda \le 4$. Equation (1) can be used to illustrate much contemporary pioneering
work: the period doubling to chaos publicized by Robert May,
the universal numbers and renormalization theory of Mitchell
Feigenbaum, and the period three implies chaos of Tien-Yien Li
and James Yorke. As λ varies, Eq. (1) exhibits a wondrously
discontinuous alternation between types of order and chaos,
but of greatest interest to us here is the full turbulence
on the unit interval which occurs at $\lambda = 4$. In order to make
this fully developed chaos transparently obvious, let us follow Stanislaw Ulam and John von Neumann in changing the dependent variable via the equation

$$Y_n = \sin^2 \pi X_n . \tag{2}$$

Simple algebra then shows that, at $\lambda = 4$, Eq. (1) is equivalent to

$$X_{n+1} = 2X_n \ (\text{mod } 1), \tag{3}$$

where (mod 1) means drop the integer part of each X_n. Like
Eq. (1), Eq. (3) maps the unit interval upon itself. Equation
(3) is a linear difference equation for which existence and
uniqueness proofs are trivial. This equation thus has a unique
solution passing through each initial iterate X_0. Indeed, the
analytic, seemingly constructive solution to Eq. (3) may be
written

$$X_n = 2^n X_0 \ (\text{mod } 1). \tag{4}$$

A more imaginative form of this solution is found by writing
X_0 as a binary digit string. For example,

$$X_0 = 0.1110000100110111... \ . \tag{5}$$

One then notes that future iterates of Eq. (3) may be obtained
simply by sequentially moving the binary point to the right in
Eq. (5) and dropping the integer part. It is absolutely
crucial to emphasize here that the obvious determinism,
existence-uniqueness of Solutions (4,5) becomes meaningful in
human terms if and only if X_0 can be specified through some
auxiliary experimental or theoretical means. But is it, in
fact, humanly possible to measure or compute the digit string
for X_0, in general? To find the answer, glance again at the
binary digit string for the X_0 of Eq. (5) but now regard this
digit string as describing a particular semi-infinite coin
toss sequence in which one means heads and zero means tails.
Viewed in this light, the set of all X_0 digit strings on the
unit inverval may be regarded as being identical with the set
of all possible random coin toss sequences. In consequence,

the X_0 digit strings are now seen to be random. Moreover, the
deterministic Eq. (3) generates a truly random sequence of X_n
iterates, a fact which follows directly from aforementioned
randomness of the X_0 digit strings; the solutions of Eq. (3)
therefore not only exist and are unique but, in general, they
are fully chaotic over the entire unit interval. But humans
can never construct these solutions because no finite
algorithm or finite experiment can compute or measure the random digit strings of most X_0. No matter how theoretically
solid Eq. (3-5) may appear to the untutored eye, they are only
formal equations containing secrets beyond human understanding.
Surely, the silvery luster of determinism, existence-uniqueness,
and formal analytic solutions is now visibly tarnished.

At this point in the discussion, an unwary reader may be
starting to form a false conclusion; be warned that it is not
merely some academic question regarding infinite digit strings
or infinite precision which is crucial in this example. Suppose we initially specify the X_0 of Eq. (4) to one binary
digit accuracy and suppose we then require only one binary
digit accuracy in all future X_n iterates. From Eq. (5), we
immediately see that we will have to add one additional binary
digit (one bit of input information) to X_0 for each additional
interate X_n (one bit of output information) we obtain. Quite
clearly here, our output information is identical to our
input information. Our algorithms, Eq. (3-5), do not compute,
they merely copy or translate input into output. Hence, we
cannot know the X_n-orbit in whole or in part unless we are
given the X_n-orbit in whole or in part beforehand. Chaos now
emerges as a mystery, a Godel-type mystery which only a god

can understand.

Let us close this section with a discussion of issues basic to the entire article. Determinism means that past and future uniquely follow from the present; in essence, determinism is a synonym for existence-uniqueness. Historically, chaos has meant irregular, erratic, disordered, or seemingly unpredictable -- an antonym for order. At a more technical level, chaos means exponentially sensitive dependence of final state upon initial state, positive Liapunov exponents, positive topological or metric entropy, fractal attractors, or the like; here chaos means that exponentially increasing knowledge of the present is required to retain meaningful contact with a receding past or an advancing future. However, the illustrative example of this section introduces the notion that chaos means deterministic randomness -- deterministic because of existence-uniqueness and randomness because the X_n-orbits are, in general, realizations of a known random process. Moreover, all prior meanings of chaos can now be subsumed under a single definition: chaos is a synonym for randomness. Indeed, we here take this as our primary definition of chaos. In the present usage, the word "random" must not be assumed to imply only a uniform probability distributions; chaos does not exclude loaded dice. But chaos is now fully defined if and only if the word "randomness" is well-defined. In this section, we have established the randomness of digit strings for most real numbers by, in essence, taking a coin toss sequence to be our definition of randomness. This approach has great intuitive appeal (which is why we used it) but, upon reflection, the reader will recognize its flaws.

Nonetheless, many probabalists agree that defining randomness
for digit strings would completely solve the definitional
question. Thus, we have hit upon the right question but have
not adequately justified our correct answer. We now proceed
to rigorously address the question of randomness in digit
strings. In so doing, we shall be led to some of the deepest
questions in science and mathematics.

2. ALGORITHMIC COMPLEXITY THEORY

The rectangle in Fig. 1 symbolizes the allowed state space
of an arbitrary finite dynamical system. This state space is
divided into two equal cells labeled by the integers 0 and 1.
As the system evolves in time from an initial state S_0, let us,
at equally spaced moments of time, observe and mark down the
number of the cell in which the evolving state S then resides.
In this way, each system orbit can be made to generate a corresponding binary digit string. Moreover, this strategem permits us to reduce the problem of randomness in a continuous
orbit to the question of randomness in a specified binary digit
string. This, in turn, then leads us to the question of ran-
donmness or its lack in specified digit strings independent of

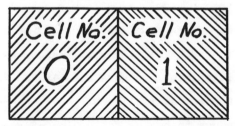

Fig. 1. - The points of this rectangle represent the states of
         a dynamical system. The division into two equal
         cells, here labeled zero and one, permits each sys-
         tem orbit to generate a binary sequence whose digits
         specify in which cell the system state resides at
         equally spaced instants of time.

their origin.   How random is a table of random numbers, for
example, or how random is the digit string in the decimal
representation of π?

   To answer these questions, let us begin by following
Andrei N. Kolmogorov, Gregory J. Chaitin, and Ray J. Solomonov
who elect to define the algorithmic complexity K_N of a specified N digit binary sequence as the number of binary digits
in the shortest computer program which prints the specified
digit string.   Basically, K_N provides a quantitative measure
of the informational compressibility or redundancy of the given
sequence; moreover, K_N may be shown to be essentially independent of specific computer.   Intuitively, a random sequence
would be expected to be incompressible while a highly ordered
sequence would contain much redundant information which could
be compressed out.   For example, consider an N digit binary
string containing only ones.   A program which can print this
sequence reads, "PRINT 1, N times."   When the number N is
sufficiently large, the size of this program is determined
almost solely by the number of binary digits required to
specify N, namely $\log_2 N$.   Thus, as an upper bound, we have
$\log_2 N \gtrsim K_N$.   The information in a string of ones is thus seen
to be logrithmically compressible.   On the other hand, any N
digit binary string can always be printed by the copy program,
"PRINT$(b_1, b_2, b_3, \ldots, b_N)$," where $(b_1, b_2, b_3, \ldots, b_N)$ is the
specified digit string.   Should this be the shortest possible
program, then we have $K_N \approx N$.   When K_N is on the order of N,
the specified digit string is so disordered, so patternlessly,
so information laden that no algorithm appreciably shorter
than the specified digit string in length can compute the

string. When K_N is appreciably less than N, the given digit string is predictably ordered, carries little information, and can be computed by a functional rule of length less than the string itself.

In consequence, Kolmogorov, Chaitin, and Solmonov define an N digit binary sequence to be random when $K_N \approx N$. This definition is ingenuous and has must intuitive appeal, but it raises several questions. Is K_N computable by a finite algorithm or have we merely replaced one undefinable by another? Is this new definition of randomness fully consistent with the older definitions? How is this definition extended to include infinite sequences? Finally, how abundant are digit sequences having $K_N \approx N$; alternatively stated, is randomness well-defined only for a null set? Let us address these questions in the inverse order of their being asked. Most binary sequences having N digits are random and have $K_N \approx N$. This follows from the paucity of short programs. There are 2^N sequences having N binary digits but only $2^{(N-k)}$ appreciably shorter programs of length (N-k) available to compute some subset of these 2^N sequences. Obviously, the subset of N digit binary sequences having complexity less than (N-k) decreases exponentially with k. In consequence, random N digit sequences are overwhelmingly abundant, but note that these arguments do not prove that any specified sequence is random. Turning now to infinite binary sequences, we elect to follow V. M. Alekseev and his Soviet colleagues who define the complexity of an infinite sequence as the limit of the normalized complexity (K_N/N) given by

$$K = \lim_{N \to \infty} [K_N/N].$$ (6)

The reason for dividing K_N by N in Eq. (6) involves a technical
point which need not concern us here; mathematically minded
readers may be assured that the limit in Eq. (6) does exist,
in general. An infinite sequence having positive complexity K
is said to be random; moreover, by the definition of complexity
such sequences cannot be computed by any finite algorithm. The
simplest way to specify them is to provide a copy. Sequences
of positive complexity are quite common; in fact, Per Martin-
Lof has proved that almost all (in the sense of measure theory)
binary strings have positive complexity. This means that the
digit strings for almost all real numbers are random and are
not computable by any finite algorithm, thus verifying the
earlier assertions made below Eq. (5). The fact that most
real numbers are not computable by any finite algorithm takes
on decidedly more funereal implications when expressed in Mark
Kac's words, "Most numbers in the continuum cannot be defined
by any finite set of words." Finally, when the digit sequence
generated by a continuous orbit is random, the orbit itself
must be random, else a random sequence of positive complexity
could be computed from a non-random oribt of null complexity.
When the sequence has null complexity, the orbit must also have
null complexity; the proof of this is simple but lengthy, and
we omit giving it.

Consistency of the algorithmic definition of randomness
with earlier definitions is verified by Martin-Lof's theorem:
Almost all infinite binary sequences having positive complexity
pass every computable test for randomness, where a computable
test is one expressible as a finite algorithm. Glancing at

Fig. 2. - Can you recognize a duck? Artist: Franco Vivaldi,
          Queen Mary College; Photographer: Vincent Mallette,
          Georgia Tech.

Fig. 2, we immediately recognize the fowl to be a wild duck
because that which swims like a duck, quacks like a duck, has
a flat bill like a duck, and colorful oily feathers like a
duck is frequently taken to be a duck. In the same vein, if
a finite sequence has $K_N \approx N$ or an infinite sequence has positive K, then the sequence is random because it "swims" like a
random sequence, "quacks" like a random sequence,... . There
is now left only one original question to answer: Is K_N
computable by a finite algorithm? In reply, let us remark
that the algorithm which computes K_N cannot itself have less
than K_N binary digits. This remark implies that K_N is, in
principle, computable as long as the number K_N stays within
humanly meaningful bounds, but alas, in the cases of greatest
interest, K_N is not bounded. As a consequence, K_N cannot be
regarded as computable, in general. The source of the problem

here is not merely the size of K_N; fundamentally, evaluating
K_N involves a problem of self-reference which is innate and
cannot be eliminated. The import of this problem for the overall theory can be appreciated from the statement: Almost all
(infinite) real number digit strings are random, but not a
single one can be proved to be random. But if K_N is uncomputable, in general, does this mean that our transition chaos
to randomness to complexity has merely been an exercise in
juggling undefinables? Hardly, regardless of whether or not
K_N is computable, recognizing that chaos means randomness
brings new unifying clarity to a previously vague and multiply
defined notion; perceiving that randomness can be defined in
terms of information and algorithmic computability again
brings a new unifying clarity to a previously vague and multiply defined notion. Nonetheless, how can we justify founding
algorithmic complexity theory upon an uncomputable base?
First, in some cases the precise K_N can perhaps be computed.
Second, in a much broader variety of cases, an easily obtained,
very approximate K_N is frequently found to be adequate. Third,
in no case is a precise value for K_N actually required for
applications. Last, but far from least, the uncomputability
of K_N, as mentioned above, is not a matter of theoretical
mismanagement but is rather a rigorous consequence of innate
human limitation. Thus, as we have sharpened chaos into
randomness into complexity, we have, without noticing it,
been carried down to the lowermost basement level of human
knowledge where the most difficult questions of self-reference
are asked by systems seeking to gain information about themselves. Let us digress briefly to consider this point.

As children, we pondered confusing puzzles of self-reference such as "A barber shaves everyone not shaving himself; who shaves the barber?" and "Does a Cretan lie when he asserts, 'All Cretans are liars.'?" Later, upon maturing, we put away childish things. Not so Kurt Godel and Alan Turing who transformed these puzzles of self-reference into powerful mathematical theorems, "This theorem is unprovable" and "This computer program cannot be proven to halt." Once proved, these theorems shook the mathematical community like an earthquake, magnitude twenty on the Richter scale. Seismic waves spread throughout even the scientific hinterlands. Subsequently, however, when no serious theoretical damage was reported, physical scientists, like the maturing children before them, put away "childish things." But alas, despite all its illusions of safety, physical science has for centuries stood astride a theoretical fault of self-reference so long and so deep that, when the tectonic plates finally move, the resulting earthquake will totally obliterate the theoretical landscape so familiar to us all.

To understand this matter, consider Fig. 3 which schematically depicts our local universe plus an "outside" clock and meter stick for making measurements. The eyes denote an observer who is a part of the local universe. Now dynamics, classical or quantal, asserts that the time evolution of the local universe can, in principle, be deterministically computed once the initial state S_0, classical or quantal, is specified. But who specifies S_0? Why is its measurement regarded as such a trivial matter? Clearly, dynamics ignores the problem of chaos discussed around Eq. (3-5), but worse,

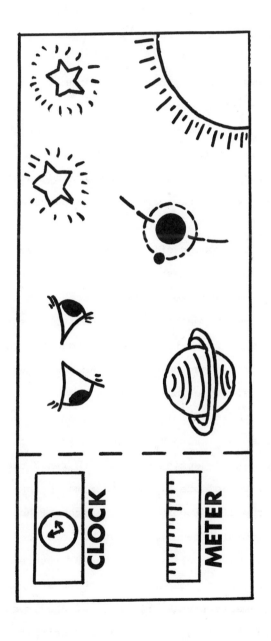

Fig. 3. – An imaginative view of our local universe showing
the clock and meter stick with which measurements
are made. The eyes denote the observer who is part
of the local universe; the dotted lines indicate
that the measuring devices lie "outside" the local
universe and interact only weakly with it.

it ignores an obvious problem of self-reference. Dynamics
assumes that S_0 can be determined from the "outside" by the
sufficiently weakly interaction or controllably interacting
clocks and meter sticks. But "outside" is a myth. In Fig. 3
there is only a composite system -- local universe plus measuring devices -- asking questions about itself. At this
point, those who ignore Godel and Turing do so at their peril.
Nonetheless, Godel's theorem becomes recognizably crucial only
when we appreciate the ubiquity of chaos, randomness, and
positive complexity. In Fig. 3, the local universe, the clock
and the meter stick are all chaotic; here, weak interactions
cause exponentially growing effects. Each weakly interaction
element of this composite system induces an uncontrollable,
unpredictable noise into the others. Precise measurement of
any variable becomes impossible. Eventually, this induced
noise may perhaps be found to have an irreducible component,
but nothing is certain at the moment. We now return to the
problem of self-reference inherent to K_N.

Let us be given an N digit binary sequence B, and let us
inquire as to its complexity K_N. Suppose it is possible to
find an algorithm A, depending on B, which finds the provably
shortest program to compute B. But this means that the search
algorithm A is capable of printing sequences B; hence, by
definition, the complexity of A cannot be less than that of B.
This self-reference, B implies A implies B, insures that
$K_A \geq K_N$, where K_A is the complexity of the search algorithm A.
However, these remarks reveal only part of the problem. Suppose we first quickly print the sequence B using the copy program just to get a feel for the time required. Then, in steps,
let us reduce program length and observe the time it takes

each successive program to print B. In general, as we
decrease program length toward K_N from above, we must antici-
pate encountering a program which does not halt during any
time interval available to us. Indeed, we know the program
will run forever if its length is less than K_N. Thus, we are
led to ask if there exist theoretical means to determine
whether or not a given program will accomplish its task and
half. Turing's version of Godel's theorem asserts that this
question is undecidable. The only way to find out whether or
not a program will stop, in general, is to let it run to
termination, if any. It is therefore no longer reasonable to
regard K_N as algorithmically computable, whether it be large
or small. In consequence, let us again discuss the basic
question, "How is complexity theory to be resuscitated from
this seemingly lethal blow?"

For the mathematical development of complexity theory,
only the existence of K_N is usually needed, not its precise
value. Equally, in many applications only some crude estimate
of K_N, not its exact value, is satisfactory. For example, if
we are given a binary sequence of N digits coming from an
arbitrary source or from the solution to an equation of motion
and if we can find a program of length appreciably less than
N which computes the given sequence, then we know this sequence
is not random; moreover, we have an upper bound on K_N. In
this circumstance, the exact value of K_N is of no great
interest. A simple case for which this occurs, as mentioned
earlier, is the digit string consisting of N ones. To print
this sequence, we first write a computer algorithm composed of
k binary digits which prints one digit of the sequence; we

then iterate this algorithm N times. In consequence, we have
$K_N \leq \log_2 N + k$. Generalizing, consider the binary representation for any rational number. The resulting digit string
periodically repeats some initial block of digits indefinitely.
Let the computer algorithm which computes the first block of
digits have length C, then the complexity of computing N
blocks is $K_N \leq \log_2 N + C$, independent of whether or not the
first block is a random digit string. Generalizing even
further, consider the first N digit segment of the binary
representation for any algebraic irrational like $2^{1/2}$ or any
transcendental irrational like e or π. All these numbers may
be computed through repeated iteration of some finite program
needing d digits to compute each output digit. Here again,
$K_N \leq \log_2 N + d$. Using Eq. (6), we immediately find that as
N → ∞ all these digit strings, including e and π, have null
complexity. Hence, the digits for e and π are not random.

Complexity theory here offers us our first mild surprise.
Investigators have long thought the digit sequences for e and
π were random. What deceived them; where or what is the order
in these digital representations? The answer hinges on the fact
that a digit string is random only if it passes all computable
tests for randomness. However in applications, this point can
be more subtle than it appears. To illustrate, consider the
base ten digit string specifying Champernowne's number

C = 0.12345678910111213141516171819202122232425...

constructed by writing the positive integers in order. The number C has been shown to be a transcendental irrational number;
moreover, Champernowne has proved that C is a normal number

which means each digit $0,1,\ldots,9$ appears with frequency 10^{-1}
and each digit sequence $d_1 d_2 d_3 \ldots d_k$ appears with frequency 10^{-k}.
The normality of the number C insures that its digit sequence
passes so many tests for randomness that even experienced
investigators sometimes assume many means all and normal means
random. However, the transparent order in the construction of
C makes it clear even to a novice that normality and randomness
are quite distinct concepts. Indeed, the digit string for C has
null complexity and it does not pass some rather obvious tests
for randomness. But let us now illustrate these same facts for
the less transparently ordered base ten sequence

$\sqrt{2}$ = 1.41421356237309504880168872...

The square root of two is an algebraic irrational number which
all numerical evidence indicates is normal; however, its digit
string has null complexity. So, where is the order in this
sequence of null complexity? As with the number C, the order
in the $\sqrt{2}$ digit string lies in its simple pattern of formation,
i.e., in its algorithmic order. The sequence for $\sqrt{2}$ may be
constructed using the iterative algorithm $A_{n+1} = [(A_n^2 + 2)/2A_n]$,
where A_n is the nth approximate to $\sqrt{2}$. Now let $A_0 = 7/5 = 1.4$;
then we find $A_1 = 99/70 = 1.414$ and $A_2 = 137,207/97,020 =$
1.41421356, where each decimal approximate A_n has truncated
according to the error in $[2 - (P_n/Q_n)^2]$. Here we note that
knowing the first two digits in the $\sqrt{2}$ sequence uniquely determines the next two digits while knowing the first four digits
uniquely determines the next five digits. Thus, the digit
string for $\sqrt{2}$ is clearly not random for the same reason that
the digit string for C was not random. Similar arguments for

e and π can now be supplied by the reader. Indeed, now that
he knows where the order resides, the reader will have little
trouble devising randomness tests not passed by C, $\sqrt{2}$, e, or
π. As a final exercise for the reader, consider the base ten
digit string for the irrational number P = 0.23571113172329...
constructed by retaining those integers in the number C which
are divisible only by unity and themselves. Since this sequence
has null complexity, find an algorithmic rule simpler than a
search through all possibilities which constructs P. We now
turn to another surprise.

Of greater interest to physical science is the question
of order or chaos in given equations of motion. As an example,
consider first

$$X_{n+1} = X_n + \omega \ (\text{mod } 1), \tag{7}$$

where ω is a given irrational number. Equation (7) has the
general solution

$$X_n = n\omega + X_0 \ (\text{mod } 1). \tag{8}$$

Since the X_0 digit strings in Eq. (8) are just as random as
they were in Eq. (4, 5), surely Eq. (8) must also describe
chaos. Let us proceed to find out. Note that first, if we
specify the error in a quantity being computed (mod 1), we are
also thereby specifying the number of its significant digits.
This is quite useful when estimating complexity since we may
replace number of input binary digits, number of output binary
digits by statements of accuracy. In the following, we work
in inverse order to the usual dynamics approach. Specifically,
we shall take the output iterates X_n (or their accuracy) as
the given, and thence we work backward via error analysis to

the amount of input X_0, ω information needed. Turning to
details, let the accuracy of each output iterate X_n, $n > 0$,
be 2^{-k+1}. For consistency in Eq. (8), we let the accuracy
of X_0 and $(n\omega)$ each be 2^{-k}. Now, if the accuracy of ω itself
is $2^{-\ell}$, then consistency with the accuracy of $(n\omega)$ requires
that $2^{-k} = n2^{-\ell}$. Then taking $\log_2$ of both sides, we find

$$\ell = \log_2 n + k. \tag{9}$$

From Eq. (9), we see that an output sequence of approximately
(nk) binary digits can be obtained for the approximate input
price of k digits for X_0 and the $\ell = \log_2 n + k$ digits for ω.
Thus, the computer program which prints the (nk) digits for
the set $(X_1, X_2, \ldots, X_n)$ requires C digits to incorporate Eq.
(8), print statements, and the like, plus k digits for X_0,
$(\log_2 n + k)$ digits for ω, and $\log_2 n$ digits for n itself.
Thus, we have $K_{nk} \leq 2(\log_2 n + k) + C$. Solution (8) therefore
has null complexity, is not chaotic, and is not random. Not
only does the randomness of the X_0 digit string not appear as
randomness in the X_n orbit, but the number of digits in X_0 is
not required to increase, when computing fixed accuracy X_n,
even as $n \to \infty$. It is the digits of ω which must increase, but
only as $\log_2 n$. Equations (7, 8) thus decline the offer of
randomness made by the continuum. Nonetheless, Mapping (7)
is ergodic, meaning that the iterates X_n, in general, densely
fill and uniformly cover the unit interval. While it is a mild
surprise to find that ergodicity does not imply randomness as
frequently assumed by equilibrium statistical mechanics, a much
deeper surprise, to which we soon return, lies in the fact that
ergodicity is not appreciably more complex, informationally

speaking, than the solvable problems of classical dynamics
such as the Kepler problem.

As our final example here, let us verify that $K_N \approx N$ for
most solutions

$$X_n = 2^n X_0 \ (\text{mod } 1) \tag{10}$$

of the difference equation in Eq. (3) using error analysis.
Here again, let all output X_n have the same accuracy 2^{-k}, and
let X_0 have accuracy $2^{-\ell}$. Then consistency in Eq. (10)
requires $2^{-k} = 2^n 2^{-\ell}$. Thence,

$$\ell = n + k. \tag{11}$$

Equation (11) asserts that, to obtain (nk) digits of the output solution sequence using Eq. (10), we must use a program of
length no less than $\ell = (n + k)$, which estimates complexity
to be on the order of n. But is this the minimum program?
Equation (10) is the unique solution; thus complexity cannot
be reduced there. The only hope remaining to reduce the overall program is to introduce a subprogram of length less than
(n + k) which computes the (n + k) digits of X_0. But this is
impossible, in general, since almost all X_0 digit strings are
random as announced by Martin-Lof. Thus, we can now assert
that each orbit of Eq. (10) has $K_N \approx N$, probability one. The
modifier, probability one, implies both a minor defect and a
major virtue. We cannot determine the precise value of complexity K_N for any specific Eq. (10) orbit, but this is only a
minor flaw. Physical science is interested in the general or
generic behavior which persists under small perturbations,
noise, or imperfect knowledge of any sort. In consequence,

knowledge that almost all Eq. (10) orbits have $K_N \approx N$ is of a
much more profound sort than is knowledge of any specific
orbit complexity. Moreover, D. V. Anosov has proved, in a
different context, that small perturbations of Eq. (10) will
not destroy the property $K_N \approx N$, probability one. Finally, in
Mapping (3) as opposed to Mapping (7), we observe the direct
transformation of randomness in the X_0 digit strings into the
randomness of the X_n orbit.

These first few sections of our article have attacked,
fang and claw, the issue of meaningfulness in the mathematical
equations used to model physical phenomena. Although specific
attack has thus far been mounted only against the discrete
logistic equation (at $\lambda = 4$), the success of that campaign
emboldens us to attack the very heartland fortresses of
theoretical physics -- Newton's equations, Maxwell's equations,
and Schrödinger's equation. For algorithmic complexity theory
is a brilliantly burnished, sharply honed sword which, if used
with wit, skill, and luck, may permit the conquering of an old
world and the creation of a new one. But to an uninitiated
reader, these first few sections may nonetheless appear to be
little more than a tempest in a teapot, merely another rummage
through the dusty old foundational closets of science. Even
if one grants that algorithmic complexity does clarify the
meaning of chaos and randomness, what strikingly new and
highly significant results flow from complexity theory regarding
meaningfulness in the well-known equations of motion? Most
especially what light does it shed on solving the unsolvable,
predicting the unpredictable? The next three sections provide
encouraging answers to these questions.

3. ALGORITHMIC INTEGRABILITY

In this section and the next, we show that conservative
Newtonian dynamics can be taken as a paradigm for illustrating
the application and usefulness of algorithmic complexity theory.
Following this, generalization to other equations or theories
can safely be left to the reader.

The question of whether or not the differential equations
of classical dynamics can be integrated to yield exact and
meaningful solutions has, over the centuries, been a vexing
one. In Newton's lifetime for example, not even the existence
or uniqueness of solutions was firmly established. But then,
concurrent with the arrival of existence-uniqueness came the
notion that all sufficiently smooth Newtonian systems were
exactly and meaningfully solvable. In consequence, one no
longer spoke of unsolvable systems, only those not yet solved.
Yet, despite all this unbridled optimism, two hundred years of
searching has revealed only one large class of exactly solvable systems, the so-called integrable systems which are the
source of almost all our textbook knowledge of classical
mechanics. An integrable system is defined to be one whose
differential equations of motion either have the form
$dP_k/dt = 0$, $dQ_k/dt = \omega_k(P_\ell)$ or can be brought to this obviously
integrable form by a change of variables. Here, as usual, the
P_k denote momentum variables, the Q_k denote position variables,
the ω_k are parameters depending on the P_k only, and t denotes
the time; there are technical restrictions such as analyticity
on the allowed variable changes, but we need not dwell on them
here. The momenta P_k for integrable systems are all constant
and, for bounded system motion, the Q_k may all be regarded as

angles. Thus an orbit for a bounded integrable system for
which k = 1,2,...,N lies on a surface having the form of an
N-dimensional torus, where the P_k specify the radii of the
torus and the Q_k specify the angles on the torus. The totality
of orbits for a bounded integrable system lie on one or more
sets of nested tori, an easily visualized and satisfying picture. Nonetheless, no matter its pleasing structure, integrability is a notion having a debilitating flaw; specifically,
integrable systems are quite exceptional; metaphorically
speaking, they are as exceptional as integers on the real line.
Continuing the metaphor a bit further, Newtonian systems
exhibiting some amount of chaos are as numerous as irrational
numbers. The density of integrable systems is, therefore, much
too low to provide appreciable insight into chaos; to provide
such insight, a class of exactly solvable systems as dense as
the rationals must be found.

Moving toward that goal, let us note that all orbits of an
integrable system may be shown to have null complexity, as one
would expect. In addition, let us recall that, quite generally,
orbits of null complexity are computable by a meaningful algorithmic rule while those of positive complexity are not.
Algorithmic complexity can thus be used as a sieve to separate
the meaningfully solvable from the unsolvable. In consequence,
let us follow Bruno Eckhardt, Franco Vivaldi, and Joseph Ford
who define a system to be algorithmically integrable (A-integrable) provided all system orbits have full complexity. The set
of A-integrable systems obviously contains as a subset all integrable systems, but more important, A-integrable systems form
the largest possible set for which all system orbits may be

computed by a meaningful algorithmic rule. Immediately beyond
A-integrability lies randomness either for individual orbits
or for entire systems. In consequence, it may be shown that
A-integrable systems can approximate randomness in the same
sense that rationals can approximate irrationals. Rather than
pursue these abstractions further, let us make A-integrability
concrete by investigating billiard motion inside various plane
polygons. The virtue of plane billiards, as a class, is that
they are perhaps the simplest systems which exhibit almost all
possible dynamical behavior.

For later reference, we begin by considering a billiard
moving within a unit square. This system is obviously integrable because, when the X-Y axes are chosen parallel to the
sides of the square, the X-motion and the Y-motion are inde-
pendent and have constant absolute speeds $|V_{x0}|$ and $|V_{y0}|$
making both X and Y coordinates sawtooth functions of the time
t. Let us now reach these conclusions via a geometric route.
In Fig. 4, we have partitioned the plane into numbered unit
squares having labeled vertices and have drawn a typical
billiard orbit as a straight line, constructed according to
the following procedure. When the billiard orbit is initiated
in the lower left square 1 as shown, it soon "collides" with
the boundary of square 1. At this point, we elect to reflect
the square rather than the orbit and thence continue the
billiard orbit as a straight line into square 2. When the
orbit "collides" with the next boundary of square 2, again the
square rather than the orbit is reflected and the straight
line continues into square 3, and so on. To regain the actual
billiard orbit in the lower left square 1, we have merely to

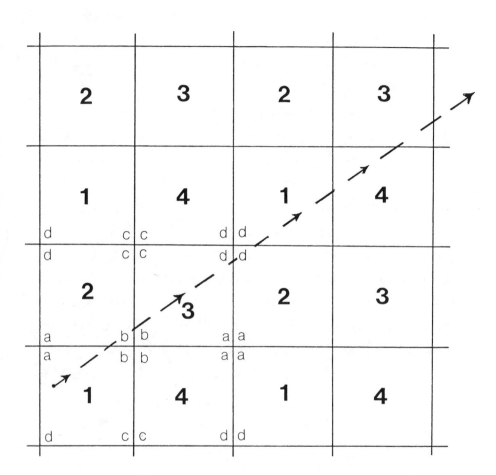

Fig. 4. - A section of the plane tiled under reflection by each
square and tiled under translation by the composite
fundamental square composed of squares 1, 2, 3, and
4. For ease of understanding, the vertices of some
squares are labeled. The dotted line represents a
typical billiard orbit on the plane.

sequentially reflect each square containing a line segment back
on top of the original square 1. Actually, this process can
be simplified. This plane is tiled not only under reflection
by a single square but also under translation by the larger
fundamental square containing squares 1, 2, 3, and 4. Thus,
if we translate all the larger fundamental squares containing
line segments back upon the lower left fundamental square, our
straight line billiard orbit has been reduced to the fundamental square shown in Fig. 5. Because of periodicity under
translation, the top and bottom sides of this fundamental
square are identical as are the left and right sides. By
physically joining top side to bottom, left side to right in
Fig. 5, we perceive that the billiard orbit is, in general,
merely a spiral on a two-dimensional torus. Integrability is
here made transparent through geometry. Analytically, the
straight line orbit on the plane is reduced to the fundamental

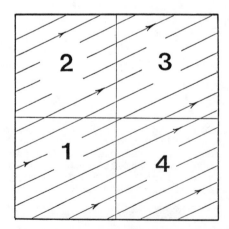

Fig. 5. - The straight line billiard orbit of Fig. 4 is here
          reduced to lie in a fundamental square. Because of
          periodicity for this fundamental square, its top
          and bottom edges are the same, as are left and right
          edges. By joining top to bottom, forming a cylinder,
          and then left to right, forming a torus, we see that
          billiard orbits for a square lie on a torus.

square via

$$X(t) = (X_0 + V_{x0}t) \ (\text{mod } 2),\tag{12a}$$

$$Y(t) = (Y_0 + V_{y0}t) \ (\text{mod } 2).\tag{12b}$$

The final reflection of orbit segments in squares 2, 3, and 4
back onto square 1, thereby obtaining the billiard orbit in
square 1, we leave as an exercise for the reader.

After this prologue, we now expose a billiard system which
is A-integrable but not integrable. Specifically, we examine
a billiard moving within a 60°-120° rhombus (parallelogram
with equal sides). As with the square, we hope to represent
the billiard orbit as a straight line on some flat surface.
This strategem will be useful only if the rhombus tiles this
flat surface under reflection and some connected group of
rhombi tile this surface under translation. In Fig. 6, we
seek to find the required surface. Figure 6ii shows rhombus
2 as the reflection of rhombus 1. Rhombus 3 is added in Fig.
6iii. If we reflect rhombus 3 on top of rhombus 1, however,
we find that vertices a and c do not match. In short, the
rhombus does not tile the plane, but planes are not the only
flat surfaces, as we shall see. In any event, let us continue
by reflecting (or rotating through 360°) rhombus 3 into a
rhombus 4 lying directly under rhombus 3. Thence, reflect
rhombus 4 into rhombus 5 lying directly under rhombus 2, and
rhombus 5 into rhombus 6 lying under rhombus 1. At last,
rhombus 6 can be reflected congruently up onto rhombus 1.
The resulting construction is pictured in Fig. 6iv where the
thick line denotes a zero width "gap" where rhombi 3-4 and 6-1
are joined. The midpoint of the double hexagon in Fig. 6iv is

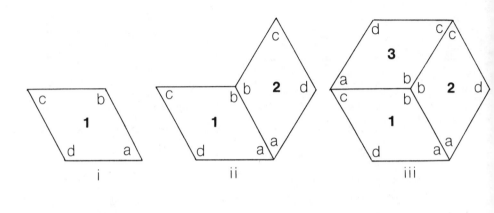

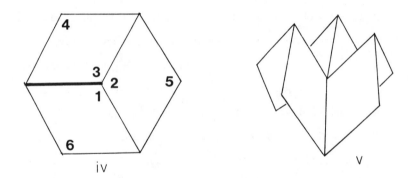

Fig. 6. - This figure shows the initial steps in the construction of the flat surface tiled under reflection by a
60°-120° rhombus. Fig. 6i, 6ii, and 6iii, show
rhombus 1 being reflected into rhombus 2 and thence
rhombus 3.   Since rhombus 3 does not properly reflect
onto rhombus 1, in Fig. 6iv, rhombus 3 has been
reflected into rhombus 4 lying directly under
rhombus 3.   Thence, rhombus 4 reflects into rhombus
5 lying under 2, and then 5 goes into 6 lying under
1.   Rhombus 6 does properly reflect up onto rhombus
1.   By opening the double hexagon of Fig. 6iv, we
find in Fig. 6v that our double hexagon is flat
everywhere (null Gaussian curvature) except at the
one saddlepoint.   In Fig. 6iv, the heavy line denotes
a zero width gap along with rhombi 3-4 and 6-1 are
joined.

a saddlepoint as is made apparent in Fig. 6v which is an
"opened up" version of Fig. 6iv. This double hexagon is thus
everywhere flat (null Gaussian curvature) except at one saddlepoint having singular negative curvature.

In Fig. 7, we seek to show the flat surface which results
from first letting the shaded 60°-120° rhombus tile the shaded
double hexagon under reflection and then letting the shaded
double hexagon tile the double plane under translation. At
each end of the heavy line segments, along which upper and
lower planes join, there is a saddlepoint. The rhombus thus
tiles under reflection a surface which is everywhere flat
except at a countable infinity of saddlepoints. Note that,
even though this almost everywhere flat surface is composed of
two joined planes, it is nonetheless but one connected surface.
As a consequence, we may draw an orbit for the billiard moving
in a rhombus as a straight line on this exotic flat surface.
A typical orbit, initiated in rhombus 1, is shown in Fig. 7.
Here the solid segments lie on the top plane while the dotted
segments lie on the bottom plane. Following this orbit closely
may aid the reader in understanding the topology of this connected double plane. This straight line orbit on the double
plane can, segment by segment, be translated back onto one
double hexagon. Then, by physically joining appropriate sides
of this periodic double hexagon, we may verify that orbits for
the rhombus lie on the two-holed doughnut surface shown in
Fig. 8. Lack of integrability for the rhombus is now made
geometrically obvious, for the rhombus orbits lie not on a
single torus as required by integrability but on a joined,
double torus. Nonetheless, there is still a disjoint link

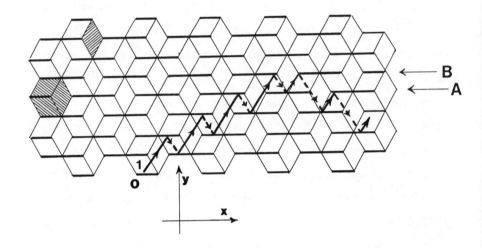

Fig. 7. - Here we see a connected double plane tiled by the
          60°-120° rhombus under reflection. To obtain this
          surface, first reflect the shaded rhombus (upper
          left) as was done in Fig. 6 to obtain the shaded
          double hexagon (middle left) and then tile the double
          plane with the double hexagon under translation. A
          typical straight line billiard orbit is drawn on this
          double plane surface with solid segments lying on the
          top plane and the dotted segments lying on the bottom
          plane. The two planes are connected along the array
          of zero width heavy line segments shown. It is to
          be emphasized that this double plane is a single con-
          nected surface which is flat everywhere except at a
          countable number of saddlepoints.

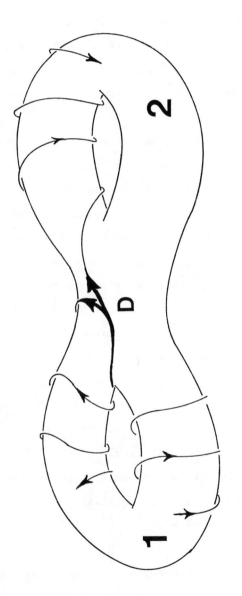

Fig. 8. – Topological representation of the periodic double
hexagon of Fig. 6iv after appropriate sides are
joined. Billiard orbits for the 60°-120° rhombus
spiral around on one or another of these two connected
tori. Orbital transition between tori are made in
a decision region here labeled D. One may show that
this decision is based on the toss of an ergodic,
but not random, coin. In consequence, these billiard
orbits have null complexity.

with integrability. Figure 8 makes it clear that each rhombus
orbit consists of connected left-torus and right-torus integrable segments. Specifically, in some neighborhood of the
decision point D in Fig. 8, the orbit must decide whether to
remain on the same integrable segment (same torus) or to "hop"
to the other. How is this decision made? Does the decision
equation have null or positive complexity; is it computable
or random?

The question of "hopping" between tori in Fig. 8 may be
shown to be identical to the question of "hopping" between
planes in Fig. 7. A billiard orbit changes planes in Fig. 8
only when it "collides" with one of the thick horizontal line
segments periodically arrayed along equally spaced lines
parallel to the X-axis of Fig. 7. Let us now investigate when
and how these "collisions" occur. First, note that, if the
60°-120° rhombus has sides of unit length, the thick horizontal line segments then have length two while the gaps between
them have unit length. Now if the billiard has reached the
horizontal level indicated by the letter A in Fig. 7, then it
will stay on the same plane if the integer part of X (mod 3)
equals zero and will change planes if the integer part of
X (mod 3) is one or two, where the origin of the X-axis
coincides with the origin of the X-Y axis system shown in
Figure 7. On the other hand, if the billiard has instead
reached the level marked B in Fig. 7, the billiard remains on
the same plane if the integer part of [X - (3/2)] (mod 3) is
zero and will change planes otherwise. Turning now from geometry to dynamics, because the thick line segments are parallel
to the X-axis, the X-motion is independent of the Y-motion,

has constant velocity V_{x0}, and may be written $X = X_0 + V_{x0}t$.
The Y-motion has constant absolute velocity $|V_{y0}|$; thus, the
billiard periodically passes through thick line segment levels
with period $T = d/|V_{y0}|$, where $d = 3^{1/2}/2$ is the Y-distance
between levels A and B. Finally then, the decision equation
for "hopping" between planes or tori is given by

$$X_n = X_0 + V_{x0}nT \pmod 3,\tag{13}$$

where no significant generality is lost by neglecting the distinction between levels A and B and their counterparts along
the Y-axis and by ignoring the value of Y_0. At last, we can
note that the decision Eq. (13) is essentially the same as
Eq. (8). Thus, one has only to repeat the error analysis previously given to establish that Eq. (13) has null complexity.
Because the integrable segments of the "60°-120°" billiard
orbits have null complexity as does the "hopping" generated by
Eq. (13), the full billiard orbit also has null complexity and
can, therefore, be expressed as a meaningful functional rule.
A billiard moving within a 60°-120° rhombus is thus A-integrable
but not integrable.

The methods outlined here to solve for billiard motion in
a rhombus appear straightforwardly generalizable to billiards
moving within any other polygon having interior angles all of
which are rational multiples of π. Although explicit solutions
have not yet been obtained for most "rational" billiards, a
theorem due to Yascha Sinai implies that billiard motion within any "rational" polygon has null complexity and, therefore,
that meaningful solutions can be found. The virtue of these
solvable "rational" billiards resides in the fact that they are
dense among all plane billiards having arbitrary boundaries.

But since, as mentioned earlier, plane billiards exhibit
almost all possible dynamical behavior including chaos,
"rational" billiards offer an analytic route to chaos paved
with null complexity; however, honesty compels us to admit
that this route becomes increasingly harder to travel the
nearer to chaos it comes.

In this section, null algorithmic complexity has been used
as a sieve to define a class of exactly solvable systems much
broader than any previously known to dynamics. Complexity
theory, of course, cannot supply the exact solution for any
A-integrable system; nonetheless, it can verify whether or not
a solution, once obtained, has null complexity. But most
important, the set of all A-integrable systems is, in a sense,
dense in the collection of all Newtonian systems. By confining
its attention to that which is possible, complexity theory has
rendered all dynamics susceptible to meaningful approximation.

4.   ALGORITHMIC RANDOMNESS

In this section, we at last directly confront the impossible, seeking to solve the unsolvable, predict the unpredictable. In particular, we here show by example how one may
analyze chaotic systems for which all or nearly all orbits
have positive complexity.

In order to physically motivate our analysis, let us very
briefly review the long-standing search for a first principles
derivation of diffusive energy transport (heat flow) in
electrically insulating solids. Note, in this regard, that
neither phenomenological nor fundamental transport theory can
provide us with a trustworthy dynamical candidate for the

study of energy transport as heat. Nonetheless, diffusive
energy transport is known to obey the phenomenological Fourier
heat law, a simple diffusion equation representing the continuum limit of a discrete random walk. In consequence, randomness is seen to be an essential ingredient of diffusive
energy transport. And now we perceive how a first principles
derivation has managed to elude a dedicated search lasting for
over eighty years. At the dynamics level, this problem
requires chaos, randomness, positive complexity, solving the
unsolvable, and predicting the unpredictable. In order to
compute energy transport or, equivalently, a thermal conductivity for a chaotic Newtonian system, we must try to compute
orbits which, like the solutions given by Eq. (4, 5), cannot
be known in whole or in part until they are given in whole or
in part beforehand. Yet, a first principles derivation of the
thermal conductivity for a particular chaotic Newtonian system
has, in fact, recently been discovered. But how can this be;
what sleight of hand, what arrangement of mirrors brought it
to pass?

Actually, it required no magic, only the replacing of that
which cannot be done by that which can. If a positive complexity value or a chaotic orbit is impossible to compute, then
let us not waste time trying to compute it; let us instead
use one or another of the generic properties of chaos to cir-
cumvent the difficulty. For example, almost all the solutions
of Eq. (10) were shown to have $K_N \approx N$ despite the impossibility
of computing K_N for any individual orbit. Similarly, although
the positive complexity of individual chaotic orbits insures
their uncomputability, this same positive complexity also

insures their randomness. A closely interwoven collection of
such random orbits would resemble nothing so much as an
insanely mixed up plate of spaghetti noodles. In consequence,
the incredibly rich variety of orbital behavior exhibited by
chaotic systems supports the intuition that any reasonable
guess for an orbit might be closely approximated by some true
orbit of the system, without, of course, providing knowledge
of which true orbit. To illustrate, the binary sequences (in
the sense of Fig. 1) generated by the solutions of Eq. (3)
are precisely the X_0 initial condition digit strings for these
solutions, as a moment's reflection will reveal. Moreover,
orbits and binary sequences are here one to one, with all possible sequences occurring. Thus, the binary sequences generated
by this chaotic system are so fabulously varied that some true
sequence approximates any finite binary sequence we arbitrarily
select as our reasonable guess; however, we can never precisely know which true infinite sequence does the approximating.
For this particular chaotic system, all possibilities occur;
however, we cannot count on this always being the case.

Turning now to a more practical situation, suppose we wish
to determine a choatic Newtonian system orbit passing through
some small ε-ball in phase space. The true system orbits
passing through this ε-ball turbulently swirl out from this
small region like smoke explosively released from a container.
While this orbital behavior is certainly quite wildly varied,
it is not so unfettered that an arbitrary orbital guess always
has a true orbit in its neighborhood. A more reasoned guess
is needed and, for this purpose, a computer is an absolute
necessity. In seeking this more reasonable guess, let us

carefully integrate numerically a chaotic Newtonian system
orbit starting from some given initial state. The resulting
numerical orbit is certainly not the true orbit passing through
the given initial state; however, this numerical orbit nonetheless constitutes a quite reasonable orbital guess, for accuracy
of numerical integration insures that, locally, each small segment of numerical orbit is close to some true orbit segment.
Moreover, qualitatively such orbital calculations cannot be
substantially improved, for this approximate orbit is close to
the best that computability can provide. Even so, our integrated orbit is only locally accurate. But now the miracle of
chaos, the exponential separation of almost all close true orbit
pairs, makes our rather crude orbital guess quite accurate
indeed. We cannot follow a true orbit, but a true orbit can
follow us. Alternatively stated, our computer calculations
has carried us straight along a path of null complexity all the
way to the rainbowed edge of chaos in whose multicolored light
our carefully integrated numerical orbit is seen to have in its
shadow some true orbit of the system. Although no human can
every actually "see" a true orbit having positive complexity,
as a recompense the gods of Chaos have benevolently provided
true shadow orbits to validate our best numerical efforts.
It is in this gift from the gods that permits man to compute
the uncomputable, predict the unpredictable. These remarks
find rigorous support in the so-called β-shadow theorem of
D. V. Anosov and Rufus Bowen. This theorem is valid for
chaotic orbital regions throughout which all initially close
orbit pairs locally separate exponentially at a rate having a
lower bound. Systems having only repulsive interparticle

forces are expected to fall under the β-shadow theorem. On
the other hand, systems having an attractive part to their
interparticle forces can, in principle, have exponentially
small stable regions in every neighborhood. However, such
microscopic stability is not expected to be a common occurrence;
nonetheless, some caution must be exercised when invoking the
Anosov-Bowen theorem. We now turn to selecting a chaotic
dynamical system suitable for the study of diffusive energy
transport.

Any dynamical candidate for validating the Fourier heat
law must meet two basic requirements. First, it must be
algorithmically random in the sense that all or nearly all its
orbits have positive complexity, and second, the interparticle
forces must be sufficiently simple and the total particle number sufficiently small that numerical analysis is feasible.
An empirical consideration adds an optional third requirement.
Even laboratory solids which obey the Fourier heat law can
exhibit non-diffusive energy transport in the form of slowly
decaying coherent excitations such as soundlike pulses and
solitary waves. Moreover, energy transport in fully chaotic
dynamic models can be strongly affected by these travelling
coherent excitations when system size is made small to expedite
numerical computation. Thus, our dynamical candidate must not
only be chaotic it must also rapidly attenuate soundlike pulses
and solitary waves even when system size is quite small.
Finally, a theoretical consideration leads to an optional
fourth requirement. In order to demonstrate that higher
dimensionality is not a crucial ingredient of thermal conductivity, we require our candidate to be a one-dimensional

system.

After nearly a decade of sifting through promising candi-
dates and discarding them one by one, Giulio Casati, Joseph
Ford, William M. Visscher, and Franco Vivaldi selected the one-
dimensional dynamical system shown in Fig. 9a to be their ideal
candidate. Here we see a system of hard point particles in
which the odd numbered open circles denote free particles and
the even numbered darkened circles denote particles harmoni-
cally bound to equally spaced, fixed lattice sites. Between
hard point collisions, all bound particles oscillate with the
same frequency ω; moreover, after suitable scaling, we find
ourselves at liberty to take ω as the sole free parameter of
the problem. When ω is set equal to zero, this system becomes
the integrable hard point gas in which all energy is trans-
ported via coherent waves (solitons); however, as ω is
increased away from zero, this system undergoes a transition
to full chaos providing an opportunity for energy to be trans-
ported solely as heat. Although we have verified the existence and nature of this transition using several technical
procedures, we believe that the reader can gain greater
physical insight into the source of this chaos from the following easily understood, intuitive argument. When ω becomes
large, all bound particles oscillate quite rapidly. In consequence, in their successive collisions with the bound particles, the free particles perceive the bound particles as having
random phase. This random phase at large ω provides that
sensitive dependence of final upon initial state known to yield
full chaos. Despite the chaos, the integration algorithm for
this system could hardly be simpler. Between collisions, this

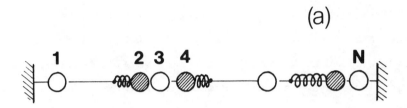

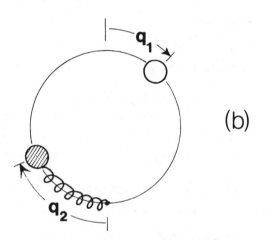

Fig. 9. - In Fig. 9a, a many-body system suitable for studying
         diffusive energy transport is exhibited. The odd
         numbered open circles denote free particles; the even
         numbered cross-hatched circles denote particles
         harmonically bound to equally spaced lattice sites.
         The springs which are drawn here are meant to ease
         visualization, but they are only symbolic and do not
         physically exist. Specifically, the particles do
         not collide with these symbolic springs. In Fig. 9b,
         we see a two particle version of Fig. 9a. Here the
         two particles move on a circle. The motion of this
         system remains ordered (integrable) as long as the
         free particle alternately collides with the left and
         right sides of the bound particle. Chaos appears
         when the free particle has multiple collisions with
         a single side of the bound particle before colliding
         again with the other side. In the fully chaotic
         hard disc gas, it is impact parameter which provides
         chaos. In this one-dimensional system, impact para-
         meter has been replaced by multiple collisions (on
         one side).

system is merely a collection of independent particles, free
and bound, whose orbits are trivial to write down. The task
of computing thus requires nothing more than keeping track of
the collision times. To illustrate the decay of solitary
pulses, suppose all system particles are initially at rest and
then suppose a large impulse is given to particle one, causing
it to move rapidly to the right in Fig. 9a. Now if we have
chosen ω to be sufficiently large (stiff spring), we can be
certain that, after the particle one-particle two collision,
bound particle two does not have the energy required to stretch
its very stiff spring far enough to hit particle three. This
coherent initial pulse has failed to propagate past even one
particle. Not all pulses or coherent excitations die this
quickly, of course, but control of ω insures control of pulse
attentuation. In summary then, the system of Fig. 9a admirably
meets all four of our requirements. Nonetheless, the proof
of a pudding still remains with the eating, so let us now sam-
ple the delicacies baked in our numerical ovens.

Diffusive energy transport obeys the Fourier heat law
which reads

$$J = -KdT/dX, \tag{14}$$

where J is the heat current (energy per unit time passing a
given point), dT/dX is the one-dimensional temperature gradient,
and K denotes an intrinsic system parameter independent of system size called the thermal conductivity. In order to validate
Eq. (14) for the system of Fig. 9a, we first simulated thermal
reservoirs at each end of the system, one hot and one cold.
When either of the free end particles crossed an arbitrarily

defined boundary, it was absorbed by the relevant reservoir and then emitted back into the system with a velocity selected according to a probability distribution characterizing the hot or cold reservoir. Using $\omega = 10.0$, a value which insured full chaos and rapid soliton decay for particle number N greater than four, we then proceeded to numerically obtain a number of orbits for this system while it interacted with the reservoirs. Following some relaxation time, each system orbit reached a steady state in which the time averaged heat current and temperature gradient were both constant. Finally, we computed the thermal conductivity K using Eq. (14) for orbits having different heat fluxes, temperature gradients, and particle number; the same conductivity value was obtained for each case. Figure 10 presents typical results; here conductivity versus particle number is plotted holding ω at the constant value of unity. Although the system is just as fully chaotic at $\omega = 1.0$ as $\omega = 10.0$, the value $\omega = 1.0$ is sufficiently low that soundlike pulses can coherently transport energy across the entire system when particle number is less than about eleven. The conductivity value stabilizes above $N = 11$. This figure clearly illustrates the transition of energy transport from sound to heat. Figure 10 also emphasizes that randomness in the heat reservoirs is not the source of diffusive energy transport.

Having validated the Fourier heat law using thermal reservoir, steady state calculations, we next sought an independent verification using the Green-Kubo linear transport theory formula for the conductivity which, very loosely speaking, derives from the diffusive spread of a localized energy packet in an isolated system. This Green-Kubo formula

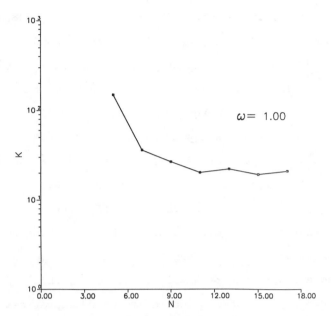

Fig. 10. - A graph of thermal conducitivity K as a function of
system particle number N. The conductivity value is
seen to stabilize for N ≳ 11.

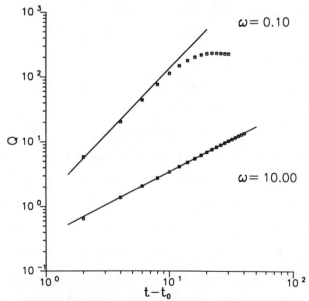

Fig. 11. - A semilog plot of a quantity Q(t), defined in the
text, versus time t. The upper curve for ω = 0.10
typifies energy transport as sound while the lower
curve for ω = 10.0 characterizes energy transport as
heat.

for the conductivity of the model in Fig. 9a can be brought to
the form $K = Q(t)/(t - t_0)$, where $Q(t)$ may be thought of as the
time varying width of an energy pulse and where t is the run-
ning system time for a calculation begun at time t_0. It is
well known that, when energy propagates as sound, $Q(t) \sim t^2$;
whereas for diffusive energy transport, $Q(t) \sim t$. In Fig. 11
we present a semilog plot of $Q(t)$ versus time for two distinct
ω-values. The slope of two for the curve $\omega = 0.10$ reveals
energy propagation as sound; the slope of unity at $\omega = 10.0$
verifies energy propagation as heat. Moreover, the conductivity obtained from the Green-Kubo formula at $\omega = 10.0$ and
$N = 48$ agrees with the reservoir value well within experimental
error. Finally, we have computed a purely theoretical conductivity value for this system using the assumption of random
phase in free-bound particle collisions. All three conductivity values are in close quantitative agreement! Our presentation of a first principles derivation of diffusive energy
transport is now complete.

In Fig. 9a, because each free particle swings back and
forth like a clapper between two harmonic oscillator bells,
we have renamed this system, originally called the Visscher
model after its inventory, the ding-a-ling model. However,
there is another reason behind the new name. Webster's dic-
tionary asserts that ding-a-ling is a euphemism for damn fool,
and a damn fool model it is indeed. This unphysical hybrid of
questionable ancestry -- half ideal gas, half harmonic
lattice -- is a true ding-a-ling for certain. Thus, many
readers may be tempted to regard the diffusive energy transport
exhibited by the ding-a-ling model as a freak occurrence in an

absurd model. But the true interpretation is quite the contrary. The ding-a-ling model has abstracted away all the seemingly relevant attributes of a proper physical system except that one unique attribute required to obtain a diffusive energy transport. Like the Cheshire cat in Alice in Wonderland, the ding-a-ling model has disappeared everything except the smile. Specifically, the bare bones of the ding-a-ling model reveals chaos to be the sole essential feature needed to obtain a diffusive energy transport. Attentuation of solitary waves or pulses is a problem only for small systems. For large systems, the transition sound to heat is concurrent with the transition order to chaos. Finally, some may object to our referring to a computer based verification as a first principles derivation. We, on the other hand, suggest that the evidence clearly shows the computer to be an absolutely indispensible tool when seeking, as here, to compute the uncomputable, predict the unpredictable. Certainly, we have not bothered to dot every "i" and to cross every "t" in our derivation, because we have no doubt it can be done. But even when this total rigor is supplied, the computer calculation will remain as the heart of this first principles derivation. The unavoidable problem here is the need for an orbit which is locally accurate everywhere permitting some true orbit to lie nearby. The only humanly feasible way to obtain this locally accurate orbit is with a computer or with some other device of the same name. Hence the computer is no longer disjoint from analysis.

5. QUANTUM CHAOS, IF ANY?

The basic question here is whether or not there is any randomness in quantum mechanics over and above the intrinsic randomness contained in the wave function ψ or the like. Such additional randomness might occur in the time evolution of the wave function or in the quantum eigenvalues and eigenfunctions. For brevity, we shall here consider only the question of randomness in the time evolution of quantal systems.

For finite particle number, spacially bounded, conservative quantum systems, it is quite well known that one can always expand the wave function $\psi(x,t)$ in the complete, orthonormal set of its energy eigenfunctions $U_n(x)$, obtaining

$$\psi(x,t) = \Sigma A_n(t) U_n(x) e^{-iE_n t/\hbar} , \qquad (15)$$

where X denotes all position variables, t denotes system time, and the $A_n(t)$ denote time dependent expansion coefficients. Putting Eq. (15) into Schrödinger's equation, $H\psi = i\hbar \partial \psi/\partial t$, we find

$$dA_n/dt + i\omega_n A_n = 0, \qquad (16)$$

where $\omega_n = E_n/\hbar$ with E_n being an energy eigenvalue. Let us now examine the question of randomness in the time evolution of the A_n, assuming the U_n-set and the E_n-set as given. Since Eq. (16) has the same form for all n, let us suppress this index and write

$$dA/dt + i\omega A = 0. \qquad (17)$$

Now into Eq. (17) substitute $A(t) = e^{-i\theta(t)}$, where $\theta(t)$ is an angle (mod 2π), yielding

$$d\theta/dt - \omega = 0. \tag{18}$$

Finally, write Eq. (18) as

$$\theta(t + \Delta t) = \theta(t) + \omega\Delta t \pmod{2\pi}, \tag{19}$$

where dt may be written Δt since t is an independent variable.
Now the method in our madness becomes clear. If we regard Eq.
(19) as providing $\theta(t+2\Delta t)$, $\theta(t+3\Delta t)$, etc., then Eq. (19) is
seen to be the same as Eq. (7). In consequence, the orbits of
Eq. (19) and hence of Eq. (16) all have null complexity. In
short, there is no randomness in the Schrödinger equation time
evolution of finite, bounded, conservative systems. But what
of infinite, unbounded, or time driven systems? Although
there is as yet no definitive proof, all existing evidence
points toward a quite general and complete lack of randomness
in Schrödinger equation time evolution.

We thus reach the paradoxical conclusion that Newtonian
systems in general exhibit randomness in their time evolution
while quantum systems do not. In particular, contrary to the
historical view, deterministic Newtonian dynamics is uncompu-
tably random while random quantum mechanics is here computably
deterministic. This paradox is heightened by the fact that
the derivation of quantal null complexity sketched around Eq.
(15-19) remains valid no matter the smallness of $\hbar > 0$; therefore, a finite, bounded, conservative quantum system cannot
exhibit any expected classical randomness even as $\hbar$ tends toward
zero. Complexity theory here places the correspondence
principle in jeopardy and perhaps even quantum mechanics itself.

The suggestion that quantum mechanics may not be random

enough will likely strike an uninitiated reader as lying somewhere between nonsense and insanity. Yet this suggestion is
now being considered with utmost gravity by both theorist and
experimentalist. Consider a hydrogenic electron in a high
Rydberg state, say n = 40, and suppose we irradiate this electron with very low energy microwave photons. Will the electron
perform a random walk up the energy ladder and escape into the
continuum or will it elect some alternative response? At the
time of this writing, Schrödinger theory predicts one thing;
semi-classical theory predicts another; which side will laboratory observation take? The battle has been joined; the conflict
is real. As cold fact without melodrama, the destiny of modern
physics may now hang in the balance.

CODA

This loving ode to algorithmic complexity theory has primarily emphasized the applications of this theory to matters
of physical or mathematical interest. Specifically, this paper
has described ways to salvage as much as possible from continuum theories which, humanly speaking, are meaningless.
But there is another direction in which complexity theory
points, alluded to briefly only here and there in the foregoing.
With even greater insistence than that provided by quantum
mechanics, complexity theory is asking us to determine the
limits of human ability and urging us to develop a theory of
natural phenomena which incorporates these limits. But in
addition to these algorithmic brave new worlds which beckon,
there remains much unfinished current business. Algorithmic
complexity theory does not consider the run time required by
its algorithms. This is a major defect because humans are just

as limited in time as in space (computer storage). Computational complexity theory, on the other hand, emphasizes the time to compute but not the space. A union of these complexity theories is devoutly to be wished. Another unfinished foundational matter is the algorithmic definition of randomness; it needs to be humanized. The fact that a sequence of positive complexity passes every computable test for randomness is beyond human need or comprehension; the humanization of this problem is not, however, a trivial exercise easily solved during an idle Sunday afternoon. Indeed, it is another algorithmic problem of self-reference which lies very deep; humans must here quantitatively determine their own limitations. An adventuresome reader undaunted by these warnings is cordially invited to spend several Sunday afternoons on this problem.

This article began with a quote from Tolstoy questioning human capacity for accepting new truths. But having now traveled to article's end, the reader can look back on new algorithmic truths which guarantee that man is capable of accepting many truths, new or old. Indeed, because humans are, in fact, capable not only of accepting but generating wondrously complex new truths, humans themselves must, of necessity, be an even more wondrously complex truth, an ancient gift to man from the primordal god whose name is Chaos. What value this gift? Hark to the answer given by the Nietzsche in his marvelously prescient benediction:

> Yea verily, I say unto you:
> A man must have Chaos yet within him
> To birth a dancing star.

REFERENCES

1. Michael Berry, "Regular and Irregular Motion," in Topics
   in Nonlinear Dynamics, AIP Conference Proceedings, Vol.
   46 (Am. Inst. Phys., New York, 1978).

2. Robert H. G. Helleman, "Self-Generated Chaotic Behavior
   in Nonlinear Mechanics," in Fundamental Problems in
   Statistical Mechanics, Vol. 5, Edited by E. G. D. Cohen
   (North-Holland, Amsterdam, 1980).

3. V. M. Alekseev and M. V. Yakobson, "Symbolic Dynamics and
   Hyperbolic Dynamical Systems," Physics Reports 75, 287
   (1981).

4. Gregory J. Chaitin, "Godel's Theorem and Information,"
   Intl. d. Theor. Phys. 22, 941 (1982).

5. Joseph Ford, "How Random is a Coin Toss?" Physics Today
   36, #4, 40 (1983).

6. Joseph Ford, "What is Chaos, That We Should be Mindful of
   It?" in The New Physics, Edited by S. Capelin and P. Davies
   (Cambridge University Press, Cambridge, 1986).

7. Bruno Eckhardt, Joseph Ford, and Franco Vivaldi, "Analytically Solvable Dynamical Systems Which Are Not Integrable,"
   Physica 13D, 339 (1984).

8. Giulio Casati, Joseph Ford, Franco Vivaldi, and William
   M. Visscher, "One-Dimensional Classical Many-Body System
   Having a Normal Thermal Conductivity," Phys. Rev. Lett.
   52, 1861 (1984).

9. Stochastic Behavior in Classical and Quantum Hamiltonian
   Systems, Lecture Notes in Physics, Vol. 93, Edited by
   Giulio Casati and Joseph Ford (Springer-Verlag, New York,
   1979).

10. Chaotic Behavior in Quantum Systems, Edited by Giulio
    Casati (Plenum Press, New York, 1985).

11. Giulio Casati, Boris V. Chirikov, and Dima L. Shepelyansky,
    "Quantum Limitations for Chaotic Excitation of the Hydrogen
    Atom in a Monochromatic Field," Phys. Rev. Lett. 53, 2525
    (1984), and references contained therein.

MAKING CHAOTIC DYNAMICAL SYSTEMS TO ORDER

Michael F. Barnsley

School of Mathematics
Georgia Institute of Technology
Atlanta, Georgia

ABSTRACT

This paper describes how strange attractors, fractal basin boundaries, ice patterns on windows, trees, ferns and classical Cantor sets can be represented to prescribed precision, as attractors for iterated function systems which can be constructed explicitly. In turn an explicit dynamical system, whose attractor approximates the given one can be exhibited.

1. INTRODUCTION

Fractals, those sets formally defined by Mandelbrot [10] to be subsets of $\mathbb{R}^n$ whose Hausdorff-Besicovitch dimensions exceed their topological dimensions, and defined in spirit more generally by the pictures in his book; when they occur either approximately or exactly, either in the observed physical world or as the output of numerical experiments, are often tell-tales of interfaces between order and chaos. Coastlines separate the churning sea and the stationary land. Intricate boundaries of basins of attraction for Newton's method, applied to seeking the zeros of a polynomial on the Riemann

sphere, carry the chaotic part of the iteration process: the
basins correspond to orderly progression towards fixed points,
while -- restricted to the boundary -- the iteration provides
an ergodic dynamical system. The apparent fractal in the ice
pattern on a window lies on the edge of the ice, between the
moving moist air and the crystal planes (poised on the pane
of glass between my warm safe bedroom and the cold buffeting
wind outside). A sign that chaos is developing as a parameter
is varied in a smooth one dimensional iterated map is the
occurrence of a bifurcation tree, containing asymptotic self-
similarities under magnification -- as described by Feigenbaum
[8] and Yorke [14] -- and thus a fractal. In the complex z-
plane the chaotic motion for iterates of $z^2 - \lambda$, as the complex
parameter λ is varied, is delineated in the complex λ-plane by
the non-self-similar arabesques of the Mandelbrot set [11, 5];
and in the study of systems of ordinary differential equations
the occurrence of strange attractors signals the presence of
chaotic dynamics. The examples are endless. More specula-
tively, a tree or a whole species of trees and their ancestors
and lineage through all time, stands to confront the universe
through all the turbulent events since the explosion of its
inception, defining an ordered stationary pattern against a
varying background; the physical boundary of the tree is the
visible interface between order and chaos.

In the studies of fractals, strange attractors and basin
boundaries one finds diverse languages. [10] refers to dusts,
dragons and curdling; Yorke to the flora and fauna of chaos;
Devaney [3] to exploding hairs and snowmen; Hubbard to
solenoids and matings; Douady to carrots and elephants; and

so on. More generally, properties of these objects are expressed in the language of analysis using various types of dimension, connectivity, accessibility, etc. However, a common language which conveys their geometry is missing: if I wish to describe my fractal to a colleague elsewhere, I had best send him a sequences of pictures since usually the diverse languages do not suffice. In some cases I can abbreviate the process. Instead of pictures I send the algorithm; for example, to convey a Julia set, send the corresponding rational map, and to study the set -- study the map. The dynamical system itself is the right language in many cases: but it will not do for the ice on my window because all I have is the pictures and no (sufficiently simple) system.

Here we address in summary an inverse problem: given a fractal interface or other intricate set, find a dynamical system whose attractor (in projection or Poincaré section) is the given set; and more generally, given a measure, find a dynamical system whose invariant measure approximates the given one. Then if the measure describes the colors of a flower the dynamical system may paint the flower in all its glorious fine detail. This does not say that the flower is actually the attractor for a dynamical system, any more than it is a set of rules used by an artist to render it. But the (hopefully) succinct description will compress the perceived information concerning the flower to something equivalent to part of what is stored in the seed, will be convenient to handle, and may be the basis of discovering new empirical laws; just as Kepler by describing planetary motion well in terms of ellipses gave Newton the neatly compressed information he needed.

2. THE COLLAGE THEOREM

The following is a simplified discussion of material in
[1, 2, 7, 9].

To approach the inverse problem we begin with the chaos
game (which I have taught twelve year old children to play
with interest). Let K be the unit square $[0,1] \times [0,1] \subset \mathbb{R}^2$,
or the screen of the graphics monitor of a microcomputer, and
let there be two continuous mappings, W_H and W_T, the 'Heads'
and 'Tails' functions, both taking K into itself.

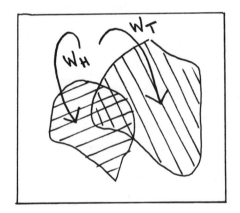

A starting point $x_0 \in K$
is chosen; a coin is tossed;
and then $x_1 =$ either $W_H(x_0)$
or $W_T(x_0)$ according as the
result is heads or tails. The
process is repeated with x_1
the new starting point; and
thus a sequence $\{x_n\}$ is
generated. If the maps are
contractions, and if all points are plotted after say fifty
iterations, they will distribute themselves approximately
upon an interesting, typically fractal set which we call the
attractor A of the iterated function system $\{K, W_H, W_T\}$. With
probability one A is the set of all accumulation points of
$\{x_n\}$. It is independent of the sequence of coin tosses, and
even of a possible bias on the coin.

As an example let K be a large piece of the Cartesian
plane which includes the interval [0,1] on the x-axis. Let

$W_H(x,y) = (x/3, y/3)$ and $W_T(x,y) = (x/3 + 2/3, y/3)$.

Then A is the classical Cantor set in [0,1] obtained by the successive removal of open middle-third subintervals: each successive iterate, whether heads or tails, is closer to [0,1]; once in [0,1] one remains there because $W_{H,T}$: [0,1]⊃ and if one is any middle third subinterval, either map takes one out of it, never to return.

More generally, let K be a compact metric space with metric $d(\cdot,\cdot)$, and let W_n: K⊃ be a contraction mapping, with for some $0 \le s_n < 1$

$$d(W_n(x),W_n(y)) \le s_n d(x,y)$$

for each n ε {1,2,...,N}. Let $p = (p_1,p_2,...,p_N)$ denote strictly positive probabilities with $\Sigma p_n = 1$.

Theorem [4].   There is a unique set A ⊂ K such that

$$A = \bigcup_{n=1}^{N} W_n(A).$$

Moreover, with probability one, A can be found by the coin toss algorithm.

Notice how A is a union of continuously altered shrunken copies of itself, which shows why A is fractal in spirit.   By the last statement in the Theorem we mean this.   Start with x_0 ε K and choose successively x_n ε {$W_1(x_{n-1}), W_2(x_{n-1}),...,$ $W_N(x_{n-1})$} with probability p_k attached to $x_n = W_k(x_{n-1})$.   Then with probability one A = {Lim x_n}.

More complete analysis of the chaos game involves con-sidering the Markov process

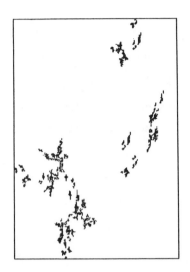

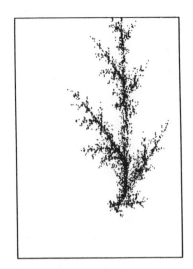

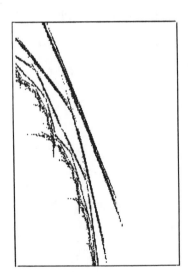

$$P(x,B) = \sum_{n=1}^{N} p_n \delta_{W_n(x)}(B),$$

where $P(x,B)$ is the probability of transfer from $x \in K$ to the
Borel subset B of K, and $\delta_y(B)$ equals one if $y \in B$ and zero
otherwise. There is a unique stationary probability measure
μ for the process, and the support of μ in A. The law of large
numbers allows us to generate μ by the coin toss algorithm.
For full discussions see [1, 4, 9, 7].

The above theorem leads one at once to wonder what different sets A one can obtain, say using three affine contractions in the plane; and it is the work of half an hour to write
a microcomputer routine to generate some pictures (below). The
results are diverse, fascinating and although one may remind
one of a cabbage and another of sealing wax, the vast majority
appear unfamiliar -- they are sets never seen in the physical
observable universe. We make this last remark because it has
application to data compression and objective recognition.
It appears that those parts of the appropriate parameter space
which correspond approximately to parts of everyday visual
images, while locally stable (nearby points give similar pictures), are few and far between. Thus, to recognize a tree
one needs to know, only roughly, where one is in parameter
space.

We are now ready for the inverse problem. Given a bounded
closed set $L \subset \mathbb{R}^2$, representing a picture say, find a collection of contraction mappings $\{W_1, W_2, \ldots, W_N\}$ whose attractor is
L. Clearly, the task is easy if L is special in the sense
that we can immediately spot how

$$L = \bigcup_{n=1}^{N} W_n(L).$$ (*)

For example, let L be the Cantor far sketched below. Clearly,

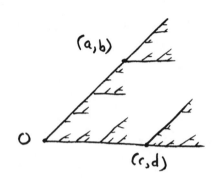

this is the union of three smaller copies of itself. If we place the origin at the main vertex and the coordinates of some other points are as shown, then (*) is true with N = 3 and

$$W_1\binom{x}{y} = \begin{pmatrix} 2/3 & 0 \\ 0 & 2/3 \end{pmatrix}\binom{x}{y} \qquad W_2\binom{x}{y} = \begin{pmatrix} 1/3 & 0 \\ 0 & 1/3 \end{pmatrix}\binom{x}{y} + \binom{a}{b}$$

$$W_3\binom{x}{y} = \begin{pmatrix} 1/3 & 0 \\ 0 & 1/3 \end{pmatrix}\binom{x}{y} + \binom{c}{d}.$$

Our problem is not with such apparently specialized sets, but rather with the approximation of any set by the attractor of a simple iterated function system.

Collage Theorem [2]. Let the contraction mappings {W_n: K: n = 1,2,...,N} be chosen such that

$$h(L, \bigcup_{n=1}^{N} W_n(L)) < \varepsilon$$

for some $\varepsilon > 0$. Then

$$h(L,A) < \varepsilon/(1-s)$$

where A is the attractor of the iterated function system.

Here $h(\cdot,\cdot)$ is the Hausdorff metric. It measures the distance between the subsets B,C of K according to

$$h(B,C) = \text{Max}\{\underset{x \in B}{\text{Max}}\ \underset{y \in C}{\text{Min}}\ d(x,y)\ ,\ \underset{y \in B}{\text{Max}}\ \underset{x \in C}{\text{Min}}\ d(x,y)\}.$$

Also, $0 \le s < 1$ is such that

$$d(W_n(x), W_n(y)) \le s \cdot d(x,y) \qquad \text{for all } x,y \in K.$$

This theorem tells us that we need only make an approximate covering or lazy tiling of L by continously distorted smaller copies of itself, in order to find a suitable set of maps. It also quantifies the continuous dependence of the attractor A of the maps $\{W_n\}$.

In the following example a leaf was approximately covered by four affine images of itself. In complex notation the maps are

$$W_n(z) = s_n z + (1-s_n) a_n$$

$$s_1 = 0.6 \qquad\qquad a_1 = 0.45 + (0.9)i$$
$$s_2 = 0.6 \qquad\qquad a_2 = 0.45 + (0.3)i$$
$$s_3 = 0.4 - (0.3)i \qquad a_3 = 0.6\ \ + (0.3)i$$
$$s_4 = 0.4 + (0.3)i \qquad a_4 = 0.3\ \ + (0.3)i$$

A sketch of the images of L under the maps, and an approximate rendering of the corresponding attractor, obtained stochastically with equal probabilities on the maps, are given next. As another example the Black Spleenwort fern (p. 60 of [13]) was considered. Four affine transformations yield the following attractor. Nearby transformations also provide recognizably fern-like images. In practice rounding error gives magnifications of the fronds an interior. Short paths in parameter space readily encode the fern trembling in the wind.

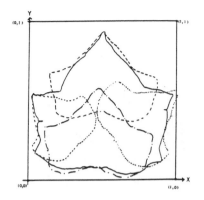

We remark on the fundamentally parallel nature of the
algorithms for regenerating the attractor of an iterated function system from its maps; this means special purpose signal
processing hardware can be designed to generate pictures at
speeds of the order of one hundred images per second.

The pictures shown here were obtained on an IBM PC with
graphics monitor. We have also generated much higher resolution pictures, with delicate shading determined by the
invariant measure, using professional computer graphics equip-
ment, see for example [6].

3. MAKING DIFFERENTIAL EQUATIONS WITH

    PRESCRIBED ATTRACTORS

It is straightforward to make a deterministic dynamical
system whose attractor (and invariant measure) project to give
the attractor (and invariant measure) of an iterated function
system. To illustrate this consider the system $\{[0,1]; W_1(x) =$
$x/3, W_2(x) = x/3 + 2/3\}$ with probability p attached to W_1 and
probability 1-p attached to W_2. Now consider the map T:
$[0,1] \times [0,1]$ defined by

$$T(x,y) = \begin{cases} (x/3,\ y/p) & \text{when } 0 \le y \le p, \\ (x/3 + 2/3),\ y/(1-p) & \text{when } p < y \le 1. \end{cases}$$

Clearly T is one-to-one and piecewise continuous. It possesses
a strange attractor and associated invariant measure whose
projections on the x-axis are the attractor and invariant measure for the original iterated function system, see also [12].

By 'slightly' changing T above one can make it continuous
while the new invariant measure is close to the original in
the appropriate weak * topology.

We now sketch out with an example how attractors and
invariant measures for iterated function systems may occur
within Poincaré sections through attractors for systems of
ordinary differential equations. We again consider the case
of the classical Cantor set. The autonomous system is

$$\dot{\underline{x}} = \underline{F}(x)$$

where $\underline{x} = (x,y,z,w)$ and $\underline{F}: \mathbb{R}^3 \times C \to \mathbb{R}^4$ is a vector field. C
is a circle of radius one, and w determines distance on the
circle. The Poincaré section which is our checkpoint for

watching the progress of orbits is a regular torus in $\mathbb{R}^3$
specified when w = 0, as sketched. In the obvious notation

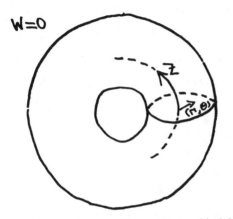

we may define it by

$$0 \leq z < 2\pi,$$
$$0 \leq \Gamma \leq 1,$$
$$0 \leq \theta < 2\pi.$$

We set $\dot{w} = 1$ so that w(t) =
t mod 2π; and the vector field
is designed to map the torus
into itself after t has
increased by 2π from 0. One
way to envisage the vector field which we are about to describe
is to think of a baker holding a big donut of compressible
dough, standing on a roundabout; as the roundabout revolves
the baker continuously deforms the dough until he is back at
the starting point (and the deformed dough is inside the part
of space occupied originally by the donut; trajectories are
traced out in four dimensions by the particles of dough. As
t (and w) increases from zero, here is what happens to the
torus. It is first sketched along its z-axis, and compressed
perpendicular to this axis to one third its radius. In the
obvious notation it is now defined by

$$0 \leq z' < 4\pi$$
$$0 \leq \Gamma' \leq 1/3$$
$$0 \leq \theta' < 2\pi$$

Finally, two very thin sections, one at $\pi-\varepsilon \leq z' \leq \pi+\varepsilon$ and
one at $3\pi-\varepsilon \leq z' \leq 3\pi+\varepsilon$ are stretched and compressed to become
hairs. The whole object is now looped to make a double torus

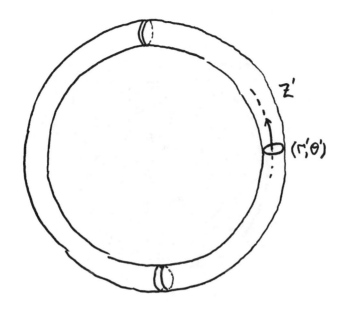

as shown, so that exactly when w = 2 it ends up inside the
original torus with its z-axis parallel to that of the original
torus except at the hairs. We call the mapping of the original
torus into itself thus defined T. The image of the disk D may
be D' = T(D) as sketched. We now consider the projection P of

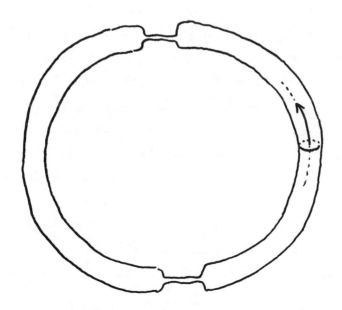

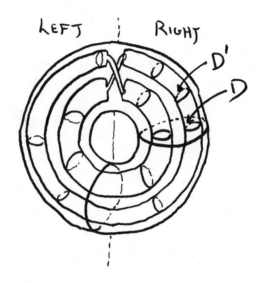

the torus, parallel to its z-axis, onto a disk K. Then for

most disks D we will have

$$P(T(D)) = \text{either } W_1(D) \text{ or } W_2(D)$$

where W_i: k are maps whose attractor is the classical Cantor

set. The only exception cases are when T(D) lies inside a

hair. Starting with almost any disk D (Lebesgue measure on

the z-axis) the sequence of its images will lie on the left

half of the time and on the right half of the time, the

sequence of lefts and rights being apparently random. If we

watch $\{P(T^n(D))\}$ it will appear that W_1 and W_2 are applied

randomly with equal probabilities, with very low positive prob-

ability (which we can make as small as we like by choosing ε

small) attached to the occurrence of a map which corresponds

to $T^n(D)$ lying inside a hair. The behavior of the iterated

function system, which now includes a collection of maps of

the latter type, is very similar to that of an iterated

function system with condensation [1]. The attractor for the sequence $\{P(T^n(D))\}$ is a web of images of the hairs, projected and mapped randomly by W_1 and W_2, accumulating on the Cantor set. The associated invariant measure has 99.999% of its mass on a prescribed neighborhood of the Cantor set, when $\varepsilon > 0$ is sufficiently small. The attractor for the differential equation, which is a solenoid, corresponding to a set of starting points of large measure, looks like a Cantor set stretched out within the torus. The dynamics within the torus, in projection, looks like the chaos game.

For an analogous case, involving four loops instead of two, the torus is like a stick of Brighton rock bent into a circle, containing in almost any cross section, to prescribed precision, images of the fern.

REFERENCES

[1]   M. F. Barnsley, S. G. Demko, Iterated function systems
      and the global construction of fractals, Proc. Roy. Soc.
      Lond. (1985) in press.

[2]   M. F. Barnsley, V. Ervin, D. Hardin, J. Lancaster,
      Solution of an inverse problem for fractals and other
      sets, Georgia Tech preprint (1984).

[3]   R. Devaney, Exploding Julia sets, this volume.

[4]   L. E. Dubins, D. A. Freedman, Invariant probabilities
      for certain Markov processes, Ann. Math. Stat. 32
      (1966), 837-848.

[5]   A. Douady, J. Hubbard, Comptes Rendus (Paris) 294 (1982),
      123-126.

[6]   S. Demko, L. Hodges, B. Naylor, Construction of fractal
      objects with iterated function systems, Proceedings of
      SIGRAPH 1985.

[7]   P. Diaconis, M. Shashahani, Products of random matrices
      and computer image generation, Stanford Univ. preprint
      (1984).

[8]   M. Feigenbaum, Quantitative universality for a class of
      nonlinear transformations, J. Stat. Phys. 19 (1978),
      25-52.

[9]   J. Hutchinson, Fractals and self-similarity, Indiana U.
      Journal of Math. 30 (1981), 713-747.

[10] B. Mandelbrot, The fractal geometry of nature, W. H.
      Freeman, San Francisco, 1982.

[11] B. Mandelbrot, Fractal aspects of the iteration
      z → λz·(1-z), Annals of N.Y. Acad. Sci. 357 (1980),
      249-259.

[12] S. Pelikan, Invariant densities for random maps of the
      interval, Trans. Am. Math. Soc. 281 (1984), 813-825.

[13] W. J. Stokoe, The observer's book of British ferns,
      Frederick Waren and Co. Ltd., London 1950.

[14] J. A. Yorke, Lectures at this conference.

ON THE EXISTENCE AND NON-EXISTENCE OF

NATURAL BOUNDARIES FOR NON-INTEGRABLE

DYNAMICAL SYSTEMS

D. Bessis*
N. Chafee

School of Mathematics
Georgia Institute of Technology
Atlanta, Georgia

ABSTRACT

We briefly review the conditions for a dynamical system to
be non-integrable. While it is generally believed that non-
integrable systems produce in the complex time plane dense
sets of singularities lying on fractals, we give arguments and
examples tending to prove that this statement is unlikely.

1. INTRODUCTION

The idea that singularities in the complex time plane may
directly influence the real time behavior of dynamical systems
is illustrated by the work of U. Frisch and R. Morf [1] and
of M. Tabor and J. Weiss [2]. In general, the singularity
nearest to the real time axis will have the greatest influence,
at least if the weight of the other singularities is not too
large. The authors just cited have recognized the existence

---

*On a leave of absence from C.E.N., Saclay, France.

Chaotic Dynamics
and Fractals

69

of an infinitely sheeted global Riemann surface associated
with the moving singularities of a non-integrable dynamical
system. A fundamental question is, on which sheet are these
singularities located? Also, can we define a Riemann distance
from such a singularity to the real time axis in such a way as
to determine the influence of that singularity on the real time
behavior? This question is not yet settled.

Another important problem is to determine the relationship,
if any between the structure of the singularities in the complex time plane and the integrability or non-integrability of
the given dynamical system. This structure might be a regular
lattice, a smooth curve, or a fractal. Also, how is this
structure related to the occurrence of chaotic behavior?

While analyzing the Henon-Heiles Hamiltonian, Y. F. Chang,
M. Tabor, and J. Weiss [3], [4] have singled out for special
attention the notion of an integrable dynamical system having
the Painlevé property, i.e., the only moving singularities of
the given system are poles. The same authors in [3], [4] have
investigated the case in which the Henon-Heiles system is non-
integrable, and in this case they have discovered that the
singularities for this system exhibit in the complex time
plane a self-similar, i.e., fractal, structure. Again, a
fundamental problem is to know on which sheet of the Riemann
surface associated with this system are these singularities
located.

Three cases may be considered. (i) The singularities
lying on the fractals are simultaneously present on all sheets
and form a natural boundary. (ii) The singularities are each
isolated in the given Riemann surface and, hence, do not form

a natural boundary. However, their projection into the complex plane produces the fractal observed. (iii) The singularities obey some combination of (i) and (ii).

In our opinion, it is important to clarify all these
questions. Our goal in this present paper is to render a
first, however, modest, contribution to that task.

In [5], the authors I. C. Percival and J. M. Green have
considered the problem of whether or not the presence of a
natural boundary is a generic property of non-integrable systems. We feel that this problem can be answered only after
the Riemann surface associated with a given non-integrable
system is better understood.

The concept of integrability varies through the literature.
J. Weiss, M. Tabor, and G. Carnevole [6] associate it with the
Painlevé property. A. Ramani, B. Dorizzi, B. Grammaticos,
and T. Bountis [7] discuss a notion of integrability under
which the Painlevé property can be absent. T. Bountis and
H. Segur [8], [9] weaken the Painlevé property by admitting
logarithmic moving singularities and introducing the concept
of weak chaos linked to them.

In [10] and [11] H. Yoshida investigates a precise but
restricted notion of integrability, namely, algebraic integrability. Yoshida gives some very interesting necessary
conditions for algebraic integrability. These conditions
involve the notion of Kowalevskaya exponents [10]. In [11]
Yoshida also produces some fractal structures for the singularities of the Henon-Heiles system, but Yoshida does not seem
to concern himself with the corresponding Riemann surface.

Finally, O. Thual and U. Frisch have studied the occurrence of natural boundaries in the Kuramoto model [14].

An overall view of all these questions is as follows.
First, one wants to link the existence of non-regular patterns
of singularities in the complex time plane to non-integrability
of the given dynamical system. One tries, in turn to link this
non-integrability with chaotic behavior. Next, one tries to
obtain an approximate description of the real time behavior of
the given system in terms of the structure of singularities
and an appropriate measure, if one can be found, on any corresponding fractal. The Fourier transform of this measure,
perhaps, can provide the insight into chaotic behavior. These
fractals, which we shall call Kowalevskaya fractals, are constructed from a knowledge of the Kowalevskaya exponents.
The construction is quite simple.

With these thoughts in mind, we suggest the following
program of research. (i) Analyze the Riemann surface corresponding to the given dynamical system. Determine its structure, at least near the real time axis. (ii) Find a functional
relation for the measure to be placed on the singularities.
(iii) Recover an approximate real time behavior from the
results of (i) and (ii).

The present paper is organized into four sections, the
first of which is this Introduction. In Section 2 we summarize
the work of H. Yoshida in [10], [11] and we review a standard
procedure for detecting singularities. In Section 3 we introduce a certain canonical example which allows one to build
arbitrary Kowalevskaya fractals. In Section 4 we discuss some
simple examples emanating from our canonical example. These

simple examples produce only isolated singularities on the
appropriate Riemann surface, but under projection onto the
complex plane those singularities give a ficticious appearance
of forming natural boundaries.

2. NONLINEAR DIFFERENTIAL EQUATIONS AND
   ALGEBRAIC INTEGRABILITY

This section concerns the work of H. Yoshida expounded in
[10] and [11].

Let us start with an example. We consider the system

$$\frac{dx_1}{dt} = \varepsilon_1 x_2 x_3$$

$$\frac{dx_2}{dt} = \varepsilon_2 x_3 x_1 \quad , \qquad\qquad (2.1)$$

$$\frac{dx_3}{dt} = \varepsilon_3 x_1 x_2$$

where $\varepsilon_1, \varepsilon_2, \varepsilon_3$ are real non-zero constants. From (2.1) we
obtain

$$\frac{x_1}{\varepsilon_1} dx_1 = \frac{x_2}{\varepsilon_2} dx_2 = \frac{x_3}{\varepsilon_3} dx_3 = x_1 x_2 x_3 dt \qquad\qquad (2.2)$$

and, hence,

$$\frac{x_1^2}{\varepsilon_1} - \frac{x_2^2}{\varepsilon_2} = c_1, \quad \frac{x_2^2}{\varepsilon_2} - \frac{x_3^2}{\varepsilon_3} = c_2, \quad \frac{x_2^2}{\varepsilon_2} - \frac{x_1^2}{\varepsilon_1} = c_3, \qquad (2.3)$$

where c_1, c_2, c_3 are constants of integration.

The functions $\phi = \phi(x_1, x_2, x_3)$ appearing in the left members of (2.3) are first integrals of (2.1). Obviously, each
one of them is functionally dependent on the other two.

From (2.3) we obtain

$$x_1^2 = \frac{\varepsilon_1}{\varepsilon_3} x_3^2 + \alpha, \qquad \alpha = -\varepsilon_1 c_3$$

$$x_2^2 = \frac{\varepsilon_2}{\varepsilon_3} x_3^2 + \beta, \qquad \beta = \varepsilon_2 c_2$$

(2.4)

Therefore,

$$x_1 = \pm \{\frac{\varepsilon_1}{\varepsilon_3} x_3^2 + \alpha\}^{1/2}$$

$$x_2 = \pm \{\frac{\varepsilon_2}{\varepsilon_3} x_3^2 + \beta\}^{1/2}$$

(2.5)

and, hence, by (2.1)

$$\frac{dx_3}{dt} = \pm \varepsilon_3 \{(\frac{\varepsilon_1}{\varepsilon_3} x_3^2 + \alpha)(\frac{\varepsilon_2}{\varepsilon_3} x_3^2 + \beta)\}^{1/2} .$$

(2.6)

It follows that, for some constant t_0,

$$t - t_0 = \pm \varepsilon_3^{-1} \int \{(\frac{\varepsilon_1}{\varepsilon_3} x_3^2 + \alpha)(\frac{\varepsilon_2}{\varepsilon_3} x_3^2 + \beta)\}^{-1/2} dx_3 .$$

(2.7)

Thus, x_1, x_2, x_3 are elliptic functions of time t.

So we see that the presence of two functionally independent first integrals in (2.3) enables one to reduce the solution of (2.1) to a problem of quadratures. Because (2.1)
possesses two independent first integrals, we say that (2.1)
is completely integrable.

We shall now consider a more general system

$$\frac{dx_i}{dt} = F_i(x_1, x_2, \ldots, x_n) \qquad (i = 1, 2, \ldots, n).$$

(2.8)

The functions F_i do not explicitly depend on the time t,
which means that (2.8) is autonomous. Therefore, if $x_i(t)$ is
a solution of (2.8), then so is $x_i(t - t_0)$ for any constant t_0.

We shall assume that the functions F_i are rational in
their arguments $x_1, x_2, \ldots, x_n$. That is

$$F_i(x_1, x_2, \ldots, x_n) = \frac{P_i(x_1, x_2, \ldots, x_n)}{Q_i(x_1, x_2, \ldots, x_n)} \qquad (2.9)$$

for $i = 1, 2, \ldots, n$, where P_i, Q_i are each polynomials in $x_1, x_2, \ldots, x_n$.

We shall also assume that (2.8) is invariant under the similarity transformations

$$t \to \alpha^{-1} t, \qquad x_i \to \alpha^{g_i} x_i \qquad (i = 1, 2, \ldots, n), \qquad (2.10)$$

where α is any nonzero constant and where $g_1, g_2, \ldots, g_n$ are fixed numbers. For example, in (2.1) above we have $g_1 = g_2 = g_3 = 1$. In general, if there exist numbers $g_1, g_2, \ldots, g_n$ such that (2.8) is invariant under (2.10) and if there exist arguments $x_1, x_2, \ldots, x_n$ at which the $n \times n$ determinant

$$\left| x_j \frac{\partial F_i}{\partial x_j} - \delta_{ij} F_i \right| \qquad (2.11)$$

is non-zero, then $g_1, g_2, \ldots, g_n$ are unique and necessarily rational.

In the presence of such invariance, we can search for a scaling or similarity solution of (2.8) having the form

$$x_i = c_i t^{-g_i} \qquad (i = 1, 2, \ldots, n), \qquad (2.12)$$

where $c_1, c_2, \ldots, c_n$ are constants, generally complex valued. Indeed, (2.12) will be a solution of (2.8) if and only if $c_1, c_2, \ldots, c_n$ satisfy the relations

$$F_i(c_1, c_2, \ldots, c_n) = -g_i c_i \qquad (i = 1, 2, \ldots, n) \qquad (2.13)$$

An important problem is to determine the structure or behavior of solutions of (2.8) near a know similarity solution (2.12). A method for doing this has been given by

S. Kowalevskaya [12]. Indeed, Kowalevskaya has shown that the
linear variational equations of (2.8) about (2.12) have solutions of the form

$$y_i = \xi_i t^{\rho - g_i} \quad (i = 1, 2, \ldots, n), \tag{2.14}$$

where ρ and $(\xi_1, \xi_2, \ldots, \xi_n)$ are any eigenvalue and eigenvector
respectively for the $n \times n$ Kowalevskaya matrix K given by

$$K_{ij} = \frac{\partial F_i}{\partial x_j} (c_1, c_2, \ldots, c_n) + \delta_{ij} g_i . \tag{2.15}$$

The eigenvalues ρ are called the Kowalevskaya exponents of
(2.12). Their arithmetical nature will play an important role
in much of what follows.

We come now to the concept of an algebraic first integral.
We shall say that a function $\Phi = \Phi(x_1, x_2, \ldots, x_n)$ is a first
integral of (2.8) if, for each solution $x_i(t)$ of (2.8), the
composite function $\Phi(x_1(t), x_2(t), \ldots, x_n(t))$ is identically
constant in t. We shall call such a first integral Φ algebraic
if, given any constant a, the relation

$$\Phi(x_1, x_2, \ldots, x_n) = a \tag{2.16}$$

can be rationalized to an algebraic equation of the form

$$a^k + \Phi_1(x_1, x_2, \ldots, x_n) a^{k-1} + \ldots$$
$$+ \Phi_k(x_1, x_2, \ldots, x_n) = 0, \tag{2.17}$$

where $\Phi_1, \Phi_2, \ldots, \Phi_k$ are rational in their arguments $x_1, x_2, \ldots,$
x_n. Given that Φ is a first integral of (2.8), the question
arises, are any of the functions $\Phi_1, \Phi_2, \ldots, \Phi_k$ first integrals
of (2.8)? The answer is given in the first of two lemmas by
Bruns (see Yoshida [10]).

Lemma 1. (Bruns) If the functions F_i in (2.8) are rational in their arguments $x_1, x_2, \ldots, x_n$, if Φ is an algebraic first integral of (2.8), and if $\Phi_1, \Phi_2, \ldots, \Phi_k$ are related to Φ as in (2.17), then $\Phi_1, \Phi_2, \ldots, \Phi_k$ are each first integrals of (2.8).

We shall say that a function $G(x_1, x_2, \ldots, x_n)$ is _weighted homogeneous_ of _weighted degree_ m, if the identity

$$G(\alpha^{g_1} x_1, \ldots, \alpha^{g_n} x_n) = \alpha^m G(x_1, \ldots, x_n) \tag{2.18}$$

holds.

Now we consider any first integral $\Phi = \Phi(x_1, x_2, \ldots, x_n)$ of (2.8) which is rational in $x_1, x_2, \ldots, x_n$. We can express Φ as a ratio of two polynomials

$$\Phi = \frac{\sum_m \Phi_m}{\sum_{m'} \Phi'_{m'}} \tag{2.19}$$

where each of the summations here is finite and where each of the elements $\Phi_m, \Phi'_{m'}$ is a weighted homogeneous polynomial in $x_1, x_2, \ldots, x_n$ with weighted degree m, m' respectively.

Lemma 2. (Bruns) Suppose that the functions F_i in (2.8) are rational in their arguments $x_1, x_2, \ldots, x_n$, and suppose that (2.8) is invariant under (2.10). If Φ is any rational first integral of (2.8) and if Φ is represented as in (2.19), then each of the rational functions $\Phi_m / \Phi'_{m'}$ is a first integral of (2.8) and is weighted homogeneous with weighted degree $m - m'$.

Thus, under our hypotheses governing (2.8), any algebraic first integral of (2.8) can be algebraically compounded from one or more weighted homogeneous rational first integrals, whose weighted degrees are rational numbers.

This brings us to the main theorems of Yoshida in [10]. In this connection, we now assume that (2.8) has a scaling solution in (2.12) with constants $c_1, c_2, \ldots, c_n)$.

Theorem 1. (Yoshida) Suppose that (2.8) has a weighted homogeneous rational first integral $\Phi = \Phi(x_1, x_2, \ldots, x_n)$ with weighted degree m. Suppose that the gradient vector

$$\nabla\Phi(\underline{c}) \equiv (\frac{\partial\Phi}{\partial x_1}(c_1, c_2, \ldots, c_n), \ldots, \frac{\partial\Phi}{\partial x_n}(c_1, c_2, \ldots, c_n))$$

(2.20)

is finite and non-zero. Then, m is a Kowalevskaya exponent for (2.12).

Theorem 2. (Yoshida) Suppose that (2.8) has two weighted homogeneous rational first integrals Φ and Φ' having the same weighted degree m. Suppose also that the two gradient vectors $\nabla\Phi(\underline{c})$ and $\nabla\Phi'(\underline{c})$ are finite and linearly independent. Then, m is a Kowalevskaya exponent for (2.12) having multiplicity at least 2.

An example illustrating all these ideas is provided by Eqs. (2.1) introduced at the beginning of this section. Indeed, the scaling exponents are $g_1 = g_2 = g_3 = 1$ and the corresponding similarity solutions are

$$x_i = c_i t^{-1} \quad (i = 1, 2, 3),$$

(2.21)

where

$$\begin{cases} c_1 = \pm(\varepsilon_2\varepsilon_3)^{-1/2} \\ c_2 = \pm(\varepsilon_3\varepsilon_1)^{-1/2} \\ c_3 = \pm(\varepsilon_1\varepsilon_2)^{-1/2} \end{cases} \qquad c_1c_2c_3 = -(\varepsilon_1\varepsilon_2\varepsilon_3)^{-1} .$$

(2.22)

Thus, there are in fact four of these similarity solutions.

The Kowalevskaya matrix is

$$
K = \begin{pmatrix} -1 & -\varepsilon_1 c_3 & -\varepsilon_1 c_2 \\ -\varepsilon_2 c_3 & -1 & -\varepsilon_2 c_1 \\ -\varepsilon_3 c_2 & -\varepsilon_3 c_1 & -1 \end{pmatrix}
\tag{2.23}
$$

and, hence, the Kowalevskaya exponents are given by

$$
\text{Det}|\rho - K| \equiv (\rho + 1)(\rho - 2)^2 = 0.
\tag{2.24}
$$

We note that these exponents ρ are independent of c_1, c_2, c_3.

From (2.3) we have two functionally independent first integrals

$$
\phi(x_1, x_2, x_3) = \varepsilon_2 x_1^2 - \varepsilon_1 x_2^2
$$
$$
\phi'(x_1, x_2, x_3) = \varepsilon_3 x_2^2 - \varepsilon_2 x_3^2 ,
\tag{2.25}
$$

and each of these is rational and weighted homogeneous with weighted degree 2. Also,

$$
\nabla\phi(\underline{c}) = (2\varepsilon_2 c_1, -2\varepsilon_1 c_2, 0)
$$
$$
\nabla\phi'(\underline{c}) = (0, 2\varepsilon_3 c_2, -2\varepsilon_2 c_3),
\tag{2.26}
$$

and these vectors are finite and linearly independent. There-fore, the number 2 must be a Kowalevskaya exponent for (2.12) having multiplicity at least 2. Of course, this last asser-tion is consistent with (2.24).

The formula (2.24) also produces a Kowalevskaya exponent $\rho = -1$. Thus number -1 is always a Kowalevskaya exponent for similarity solutions of the sort (2.12), provided that $g_i c_i \neq 0$ for at least one integer $i = 1, 2, \ldots, n$. Indeed, given that $x_i = c_i t^{-g_i}$ is a solution of (2.8), it follows that $\tilde{x}_i = c_i(t-t_0)^{-g_i}$ is also a solution of (2.8) for any constant t_0. The partial derivatives $\partial\tilde{x}_i/\partial t_0$ evaluated at $t_0 = 0$ will form a solution $y_i(t)$ of the linear variational equations of (2.8) about (2.10). Moreover, we see that

$$y_i = g_i c_i t^{-1-g_i} . \tag{2.27}$$

Comparing (2.27) with (2.14) we see that -1 must be a Kowalevskaya exponent for (2.12).

We now turn our attention to the work of Yoshida given in [11]. We continue our study of (2.8), and we continue to assume that (2.8) has a similarity solution (2.12).

Theorem 3. (Yoshida) Suppose that the Kowalevskaya matrix K associated with (2.8) and (2.12) is diagonal, and suppose that K yields k, $1 \le k \le n$, Kowalevskaya exponents $\rho_1, \rho_2, \ldots, \rho_k$ having positive real parts. If $\rho_1, \rho_2, \ldots, \rho_k$ are rationally independent, then (2.8) has a k-parameter family of solutions

$$x_i(t) = t^{-g_i}[c_i + P_i(I_{\rho_1} t^{\rho_1}, I_{\rho_2} t^{\rho_2}, \ldots, I_{\rho_k} t^{\rho_k})]$$
$$(i = 1, 2, \ldots, n), \tag{2.28}$$

where $P_i = P_i(z_1, z_2, \ldots, z_k)$ is a function analytic in $z_1, z_2, \ldots, z_k$ about the origin with $P_i(0, 0, \ldots, 0) = 0$ and where $I_{\rho_1}, I_{\rho_2}, \ldots, I_{\rho_k}$ are arbitrary constants.

We shall say that the system (2.8) is algebraically integrable if (2.8) has n-1 functionally independent algebraic first integrals.

Theorem 4. (Yoshida) If (2.8) is algebraically integrable, then every Kowalevskaya exponent ρ of (2.12) is a rational number. Therefore, if (2.12) has at least one complex or irrational Kowalevskaya exponent, then (2.8) is not algebraically integrable.

As a simple application of Theorem 4 we shall consider the planar N-body problem.

Let m_i, $\vec{q}_i$, and $\vec{p}_i$ be the mass, position, and momentum respectively of the ith particle ($i = 1, 2, \ldots, N$). We can write

$$\vec{q}_i = \begin{pmatrix} x_i \\ y_i \end{pmatrix} \qquad \vec{p}_i = \begin{pmatrix} p_{x_i} \\ p_{y_i} \end{pmatrix} \tag{2.29}$$

$$H = \sum_{j=1}^{N} \frac{|\vec{p}_j|^2}{2m_j} - \sum_{i,j=1}^{N} \frac{m_i m_j}{|\vec{q}_i - \vec{q}_j|}$$

and we have

$$\frac{d\vec{q}_i}{dt} = \frac{\partial H}{\partial \vec{p}_i}$$
$$\qquad\qquad (i = 1, 2, \ldots, N). \tag{2.30}$$
$$\frac{d\vec{p}_i}{dt} = - \frac{\partial H}{\partial \vec{q}_i}$$

To transform (2.30) into a system involving only rational functions, one introduces $(N)(N-1)/2$ new variables

$$s_{ij} = |\vec{q}_i - \vec{q}_j|^{-1} \quad (i \neq j) \tag{2.31}$$

and obtains

$$\frac{d\vec{q}_i}{dt} = \frac{\vec{p}_i}{m_i}$$

$$\frac{d\vec{p}_i}{dt} = - \sum_{j=1}^{n} m_i m_j s_i^3 (\vec{q}_i - \vec{q}_j) \qquad (2.32)$$

$$\frac{ds_{ij}}{dt} = -s_{ij}^3 (\vec{q}_i - \vec{q}_j) \cdot (\frac{\vec{p}_i}{m_i} - \frac{\vec{p}_j}{m_j}).$$

This system admits (N)(N-1)/2 first integrals

$$\gamma_{ij} = |\vec{q}_i - \vec{q}_j|^2 - \frac{1}{s_{ij}} \qquad (2.33)$$

For those solutions along which $\gamma_{ij} = 0$ the system (2.32) reduces to the system (2.30).

The system (2.32) scales in the following way:

$$t \rightarrow \alpha^{-1} t$$

$$\vec{q}_i \rightarrow \alpha^{-2/3} \vec{q}_i$$

$$\vec{p}_i \rightarrow \alpha^{1/2} \vec{p}_i \qquad (2.34)$$

$$s_{ij} \rightarrow \alpha^{2/3} s_{ij}$$

When N = 2 the system (2.32) is of order 9 and the Kowalevskaya determinant is

$$K(\rho) = (\rho + \frac{4}{3}) (\rho+1) (\rho + \frac{2}{3})^2 (\rho + \frac{1}{3}) (\rho) (\rho - \frac{1}{3})^2 (\rho - \frac{2}{3}). \qquad (2.35)$$

All the Kowalevskaya exponents are rational. As is well known, when N = 2, the system (2.32) has sufficiently many algebraic first integrals so that its solution can be reduced to quadratures.

One can easily recognize some of the algebraic first integrals associated with the exponents ρ yielded by (2.35).

For example,

$$p_{x_1} + p_{x_2} = \text{constant} \quad (\rho = \tfrac{1}{3})$$

$$p_{y_1} + p_{y_2} = \text{constant} \quad (\rho = \tfrac{1}{3})$$

$$(m_1)(x_1 p_{y_1} - y_1 p_{x_1}) + (m_2)(x_2 p_{y_2} - y_2 p_{x_2}) = \text{constant}$$
$$(\rho = -\tfrac{1}{3})$$

$$H = \text{constant} \quad (\rho = \tfrac{2}{3}) \tag{2.36}$$

When $N = 3$ the system (2.32) has 15 dependent variables, and the Kowalevskaya determinant is

$$K(\rho) = (\rho + \tfrac{4}{3})^3 (\rho + \tfrac{2}{3})^2 (\rho - \tfrac{1}{3})^2 (\rho + \tfrac{1}{3})(\rho)(\rho + 1)(\rho - \tfrac{2}{3}) \cdot$$

$$(\rho - \rho_+^{(1)})(\rho - \rho_-^{(1)})(\rho - \rho_+^{(2)})(\rho - \rho_-^{(2)}), \tag{2.37}$$

where

$$\rho_\pm^{(1)} = \tfrac{1}{6}[-1 \pm \{13 + 12\chi\}^{1/2}]$$

$$\rho_\pm^{(2)} = \tfrac{1}{6}[-1 \pm \{13 - 12\chi\}^{1/2}] \tag{2.38}$$

$$\chi = \frac{\{\tfrac{1}{2}[(m_1 - m_2)^2 + (m_2 - m_3)^2 + (m_3 - m_1)^2]\}^{1/2}}{m_1 + m_2 + m_3}$$

In general, these last four exponents $\rho_\pm^{(1)}$ and $\rho_\pm^{(2)}$ are irrational or complex, and we see that the 3-body problem is not algebraically integrable.

This completes our summary of Yoshida's work. Now we want to discuss the main question of interest to us: What is the analytical structure of singularities in the complex t-plane for integrable and non-integrable systems? In this connection, how is chaos linked with non-integrability and how is chaos reflected in the structure of those singularities?

The standard belief is that non-integrable systems produce
natural boundaries composed of singularities forming sets
dense on fractals associated in a certain way with Kowalevskaya
exponents. We shall call such fractals <u>Kowalevskaya fractals</u>.

Singularities in the complex t-plane can be classified as
either fixed or moving. By definition, the locations of
fixed singularities are independent of the initial data. On
the other hand, the locations of moving singularities vary with'
changing initial data, and, therefore, moving singularities are
difficult to study. The simplest sort of moving singularity
is a pole, and dynamical systems whose only moving singularities are poles are called systems of Painlevé type.

Now we want to discuss the detection of singularities.
The method used is the Weierstrass construction of the star
of holomorphy. This method has been computerized by Y. F.
Chang [13].

Starting from a regular point in the complex plane, the
program recursively computes a truncated Taylor series for the
given solution at this regular point. Using this truncated
series, the program then calculates a truncated Taylor series
at a new point, suitably close to the initial point, lying
on some previously selected path in the complex plane.
Having done this, the program moves on to another point, and
this process can be repeated as often as necessary. Thus,
one can detect the location and nature of singularities lying
on the circles of convergence.

If $x_i(t)$ denotes the solution being studied and if t_* is
the singularity being detected, then nearby t_* we might have

$$x_i(t) = (t - t_*)^{\alpha} H_i(t), \tag{2.39}$$

where α is some constant and $H_i(t)$ is analytic about t_*. This
is a first reasonable guess. A more sophisticated guess,
which still excludes essential singularities, is

$$x_i(t) = \sum_k (t - t_*)^{\alpha_k} H_{ik}(t), \tag{2.40}$$

where α_k are constants and where the elements $H_{ik}(t)$ are functions analytic about t_*. In general, we allow the constants
α_k to be irrational or complex.

Adhering to the simpler form (2.39) we distinguish the
following three cases. (i) α is a negative integer -p and t_*
is a pole of order p. (ii) α is a rational number p/q and t_*
is an algebraic singularity of order q which generates a q-
sheeted Riemann surface. (iii) α is an irrational or complex
number and t_* is a transcendental singularity which generates
locally an infinitely sheeted Riemann surface.

One can ask, how are these Riemann surfaces associated to
the various singularities connected to each other, and thus
what is the global nature of the corresponding Riemann surface?
This question may be important for understanding chaotic
behavior.

Many examples of obtaining the singular set occur in the
literature. In [1] the singularities in the complex t-plane
reflect "bursts" centered at the real parts of those singularities with an overall amplitude decreasing exponentially
with the imaginary part. Similar results are found in [2] for
the Lorenz system.

In [3,4] the authors describe self-similar structures for
the singularities of the Henon-Heiles Hamiltonian. They use

the concept of "canonical resonance" rather than the equivalent
concept of Kowalevskaya exponents. A good description of the
fractal structure is given.

In [8,9] the connection between the existence of logarithmic
singularities and the occurrence of weak chaos is discussed.
In [14] the Kuramoto model describing the evolution of a flame
front is investigated, and fractal structures for singularities
are found both using the Chang program and constructing the
Kowalevskaya fractal.

Here we shall briefly report an analysis rendered by
Yoshida in [11] for the Henon-Heiles Hamiltonian. This is
not essentially different from the pioneering work of Chang,
Tabor, and Weiss [3], but it is rendered in the framework
associated with Kowalevskaya exponents.

We consider the Henon-Heiles Hamiltonian

$$H = \frac{1}{2} (p_1^2 + p_2^2) + q_1^2 q_2 + \frac{\varepsilon}{3} q_2^3 , \qquad (2.41)$$

where ε is a real valued parameter.

The scaling exponents g_i corresponding the variables q_i
are equal to 2. With these exponents there are associated
four similarity solutions whose q-components are given by

$$q_i(t) = c_i(t - t_*)^{-2} \qquad (2.42)$$

with

$$c_1^{(0)} = c_2^{(0)} = 0 \qquad (2.43)$$

$$c_1^{(1)} = \pm 3\sqrt{2 - \varepsilon}, \qquad c_2^{(1)} = -3 \qquad (2.44)$$

$$c_1^{(2)} = 0, \qquad c_2^{(2)} = -\frac{6}{\varepsilon} . \qquad (2.45)$$

Ignoring the first of these solutions, one can calculate the Kowalevskaya exponents for the other three. Two of these exponents are -1 and 6. The other two are the roots $\rho_{\pm}^{(1,2)}$ of the equations

$$\rho^2 - 5\rho + (6)(2 - \varepsilon) = 0 \tag{2.46}$$

$$\rho^2 - 5\rho + (6)(1 - 2\varepsilon^{-1}) = 0 \tag{2.47}$$

corresponding to (2.44) and (2.45) respectively.

For $-48 < \varepsilon < 0$ the exponents $\rho_{\pm}^{(1,2)}$ are complex and the Henon-Heiles is not algebraically integrable. In the vicinity of t_*, the location of the double pole singularity, one can write, following equation (2.28),

$$q_i(t) = \tau^{-2}[c_i + P_i[I_+\tau^{\rho_+}, I_-\tau^{\rho_-}, I_6\tau^6]$$
$$(\tau \equiv t - t_*) \tag{2.48}$$

One can show that H is proportional to I_6 (H being a weighted first integral of weighted degree 6). Choosing $H = 0$ implies $I_6 = 0$ one obtains

$$q_i(t) = \tau^{-2}[c_i + P_i(I_+\tau^{\rho_+}, I_-\tau^{\rho_-})] \tag{2.49}$$

where ρ_- is the complex conjugate of ρ_+.

Suppose first $I_- = 0$, then

$$q_i(t) = \tau^{-2}[c_i + P_i(I_+\tau^{\rho_+})] \tag{2.50}$$

On the circle of convergence for $P_i(z)$ there is at least one singularity, say $z = z_0$. When

$$I_+(t - t_*)^{\rho_+} = z_0 \tag{2.51}$$

$q_i(t)$ is singular at

$$t_n = t_* + \left(\frac{z_0}{I_+}\right)^{1/\rho_+} e^{\frac{2i\pi n}{\rho_+}} . \tag{2.52}$$

These points lie on a spiral

$$t_n = t_* + Ms^n$$

$$s = e^{\frac{2i\pi}{\rho_+}} = re^{i\theta}$$

$$r = \exp \frac{2\pi\rho_I}{|\rho_+|^2} , \quad \theta = \frac{2\pi\rho_R}{|\rho_+|^2} \tag{2.53}$$

$$\rho_+ = \rho_R + i\rho_I .$$

where ρ_R is the real part of ρ_+ and ρ_I the imaginary part. For the solutions (2.44) we have

$$r = r_1 = \exp \frac{-\pi\sqrt{23 - 24\varepsilon}}{6(2 - \varepsilon)} , \qquad \theta = \theta_1 = 5\pi/6(2 - \varepsilon) \tag{2.54}$$

If $I_- \neq 0$ and $|I_+\tau^{\rho_+}| \gg |I_-\tau^{\rho_-}|$ then the complete set of singularities is well approximated by the spiral above. For $|I_+\tau^{\rho_+}| \ll |I_-\tau^{\rho_-}|$ we substitute $\rho_+ \to \rho_-$ and we find that $\rho = re^{i\theta}$ goes into $\rho = \frac{1}{r} e^{i\theta}$. This generates a new spiral. Hence, around a singularity with two complex conjugate exponents there are two convergent semi-spirals, one clockwise and the other counterclockwise.

For $\varepsilon = -1$ one finds $\theta_1 = 50°$ and $\theta_2 = 25°$ and this leads to an approximate set of singularities which, from two singularities, produce a third which is at the vertex of an isoceles triangle with angles 25°. This produces a fractal, the dimension of which can be easily evaluated.

An important question is: Are these singularities on the same Riemann sheet, or is the fractal just the projection onto

ℂ of isolated singularities of the Riemann surface?

We shall now produce simple examples in which all singularities are isolated, but their projection produces dense closed sets which will give the illusion of the existence of a natural boundary.

3. A CANONICAL EXAMPLE

To help us understand the structure of singularities associated with a non-integrable dynamical system, we are now going to study a certain system of differential equations.

To begin, we choose any n distinct numbers $\alpha_1, \alpha_2, \ldots, \alpha_n$. These numbers may be real or complex, but we shall require that no one of them be equal to any of the integers $0, 1, \ldots, n-2$.

Next, we introduce constants $a_1, a_2, \ldots, a_n$ by setting

$$a_n = (-1)^n \alpha_1 \alpha_2 \cdots \alpha_n$$

$$a_{n-k} = \frac{(k-\alpha_1)(k-\alpha_2)\ldots(k-\alpha_n)}{k!} - \frac{a_n}{k!}$$

$$- \frac{a_{n-1}}{(k-1)p} - \cdots - \frac{a_{n-k+1}}{1!} \quad (k = 1, 2, \ldots, n-1) \tag{3.1}$$

With these constants in hand, we now write down the following system of ordinary differential equations:

$$\frac{dx_i}{dt} = -\frac{x_i^2}{x_{i+1}} \quad (i = 1, 2, \ldots, n)$$

$$\frac{dx_n}{dt} = x_n^2 \sum_{j=1}^{n} \frac{a_j x_{n+1}^j}{x_{n-j+1}} \tag{3.2}$$

$$\frac{dx_{n+1}}{dt} = -x_{n+1}^2$$

Clearly, the right hand members of (3.2) are rational functions of $x_1, x_2, \ldots, x_n, x_{n+1}$.

The system (3.2) is invariant under the following scaling transformations:

$$t \to \alpha^{-1} t$$

$$x_i \to \alpha^{g-i+1} x_i \qquad (i = 1, 2, \ldots, n) \tag{3.3}$$

$$x_{n+1} \to \alpha x_{n+1} .$$

Here, g can be chosen arbitrarily. However, as we shall shortly see, only finitely many, properly selected, values of g yield scaling solutions of (3.2).

If to (3.2) we apply the change of variables

$$x_i = z_i^{-1} \qquad (i = 1, 2, \ldots, n, n+1), \tag{3.4}$$

the system (3.2) will become

$$\frac{dz_i}{dt} = z_{i+1} \qquad (i = 1, 2, \ldots, n-1)$$

$$\frac{dz_n}{dt} = \frac{a_n z_1}{(z_{n+1})^n} - \frac{a_{n-1} z_2}{(z_{n+1})^{n-1}} - \cdots - \frac{a_1 z_n}{z_{n+1}} \tag{3.5}$$

$$\frac{dz_{n+1}}{dt} = 1$$

Therefore

$$z_{n+1} = t - t_* \qquad t_* = \text{constant} \tag{3.6}$$

and we see that z_1 satisfies an equation of Euler type

$$(t - t_*)^n \frac{d^n z_1}{dt^n} + \sum_{j=1}^{n} a_j (t - t_*)^{n-j} \frac{d^{n-j} z_1}{dt^{n-j}} = 0. \tag{3.7}$$

The general solution of this last equation (3.7) can be written as

$$z_1(t) = \sum_{j=1}^{n} A_j (t - t_*)^{\alpha_j} , \qquad (3.8)$$

where $A_1, A_2, \ldots, A_n$ are arbitrary constants and where $\alpha_1, \alpha_2, \ldots, \alpha_n$ are exactly the numbers chosen at the beginning of this section.

From all of this it follows that (3.2) has an $(n+1)$-parameter family of solutions by

$$x_1(t) = \left\{ \sum_{j=1}^{n} A_j (t - t_*)^{\alpha_j} \right\}^{-1}$$

$$x_i(t) = \left\{ \sum_{j=1}^{n} A_j (\alpha_j)(\alpha_j - 1)\ldots(\alpha_j - i + 2)(t - t_*)^{\alpha_j - i + 1} \right\}^{-1} \qquad (3.9)$$

$$(i = 2, 3, \ldots, n)$$

$$x_{n+1}(t) = (t - t_*)^{-1}$$

where $A_1, A_2, \ldots, A_n$, t_* are $n+1$ arbitrary constants with $A_1, A_2, \ldots, A_n$ not all zero. In fact, (3.9) represents every solution of (3.2) for which $x_{n+1}(t) \not\equiv 0$.

From the same formulas (3.9) we can obtain scaling solutions of (3.2) having the general form (2.10). Indeed, for each integer $k = 1, 2, \ldots, n$, Eqs. (3.2) have a 1-parameter family of scaling solutions

$$x_1(t) = \theta_k t^{-\alpha_k}$$

$$x_i(t) = \frac{\theta_k t^{-\alpha_k - i + 1}}{(\alpha_k)(\alpha_k - 1)\ldots(\alpha_k - i + 2)} \qquad (i = 2, 3, \ldots, n) \qquad (3.10)$$

$$x_{n+1}(t) = t^{-1}$$

where θ_k is any non-zero constant. Clearly, these solutions (3.10) correspond to taking $g = \alpha_k$ in (3.3).

With the aid of (3.9) we can obtain the Kowalevskaya

exponents associated with any one of the similarity solutions
(3.10). Indeed, we can show that, for any integer
$\ell = 1,2,\ldots,n$, the number $\alpha_\ell - \alpha_k$ is a Kowalevskaya exponent
for (3.10).

Briefly put, the proof is as follows. In (3.9), for each
integer $i = 1,2,\ldots,n,n+1$, we take the partial derivative of
$x_i(t)$ with respect to A_ℓ and then set $t_* = 0$, $A_k = \theta_k^{-1}$, and
$A_j = 0$ for all $j = 1,2,\ldots,n$ such that $j \neq k$. The results of
these calculations are functions $y_i(t)$ which we can write in
the form

$$y_1(t) = n_1 t^{-\alpha_k} t^{\alpha_\ell - \alpha_k}$$

$$y_i(t) = n_i t^{i-1-\alpha_k} t^{\alpha_\ell - \alpha_k} \qquad (i = 2,3,\ldots,n) \qquad (3.11)$$

$$y_{n+1}(t) = 0$$

where $n_1, n_2, \ldots, n_n$ are non-zero constants.

By their very construction, the function $y_i(t)$ form a
solution of the linear variational equations of (3.2) about the
similarity solution of (3.10). Comparing (3.11) with (2.14),
we see that $\alpha_\ell - \alpha_k$ is a Kowalevskaya exponent of (3.10).

Thus, each of our scaling solutions (3.10) has $n+1$
Kowalevskaya exponents $\gamma_1, \gamma_2, \ldots, \gamma_n, \gamma_{n+1}$ given by the formulas

$$\gamma_\ell = \alpha_\ell - \alpha_k \qquad (\ell = 1,2,\ldots,n)$$
$$\qquad (3.12)$$
$$\gamma_{n+1} = -1$$

The presence here of the exponent -1 follows from arguments we
have already presented in connection with (2.27) in Section 2.
We can observe that in (3.12) we have $\gamma_k = 0$. In other

words, the number 0 is a Kowalevskaya exponent for each of our scaling solutions (3.10).

The real significance of (3.12) is this. If we select any n - 1 distinct numbers $\gamma_1, \gamma_2, \ldots, \gamma_{n-1}$, then we can construct a system (3.2), having order n+1, such that (3.2) will produce a scaling solution whose Kowalevskaya exponents are exactly the number $\gamma_1, \gamma_2, \ldots, \gamma_{n-1}, 0, -1$.

For this reason, we regard (3.2) as a canonical example producing, so to speak, singularities made to order. We believe that the study of (3.2) may well shed light on the general questions raised earlier in this paper, i.e., questions concerning the structure of singularities and of Riemann surfaces associated with dynamical systems.

4. SOME SIMPLE EXAMPLES

In this section we will investigate the structure of singularities coming from simple instances of our canonical example (3.2). Our work is organized under five headings 4-A through 4-E.

4-A. Let us start with the simplest situation. We can imagine a system of the type (3.2) producing a solution $x_i(t)$ whose first component $x_1(t)$ is

$$x_1(t) = \theta_1 (t-t_*)^{-\alpha_1},$$ (4.1)

where θ_1, t_*, α_1 are constants with $\theta_1 \neq 0$.

This example has a moving singularity at $t = t_*$, provided that α_1 is not a negative integer or zero. If α_1 is a positive integer, then we have a moving pole. If α_1 is rational, then

we have an algebraic moving singularity with finitely many
sheets in the Riemann surface. If α_1 is irrational or complex,
then t_* and ∞ are each logarithmic singularities, with t_*
moving and ∞ fixed. The logarithm function uniformizes $x_1(t)$.

4-B. The next example is

$$
\begin{aligned}
x_1(t) &= \{A_1(t-t_*)^{\alpha_1} + A_2(t-t_*)^{\alpha_2}\}^{-1} \\
&= \frac{A_2^{-1}(t-t_*)^{-\alpha_1}}{A_1 A_2^{-1} + (t-t_*)^{\alpha_2-\alpha_1}} \quad ,
\end{aligned}
\tag{4.2}
$$

where α_1, α_2 are chosen so that $\alpha_2 - \alpha_1$ will be irrational or
complex. Here we obtain a moving singularity at t_* of tran-
scendental type and also moving poles at

$$
t_n = t_* + (-A_1 A_2^{-1})^{\frac{1}{\alpha_2-\alpha_1}} e^{\frac{2in\pi}{\alpha_2-\alpha_1}} \quad (n = 0, \pm1, \pm2, \ldots). \tag{4.3}
$$

Setting

$$
\frac{1}{\alpha_2-\alpha_1} = \delta + i\eta \tag{4.4}
$$

we can rewrite (4.3) as

$$
t_n = t_* + (t_0 - t_*)e^{-2\pi n\eta}e^{2\pi in\delta} \quad (n = 0, \pm1, \pm2, \ldots). \tag{4.5}
$$

Suppose that $\eta = 0$ and δ is irrational. Then, in the complex plane $\mathbb{C}$ the poles t_n will form a set dense on the circle
of center t_* and radius $|t_0 - t_*|$. However, these poles do not
form a natural boundary for $x_1(t)$. This is because with $x_1(t)$
there is associated an infinitely sheeted Riemann surface
emanating from t_*, and on this Riemann surface the poles t_n
are each isolated. Within that Riemann surface the poles t_n
lie on a helix whose axis is at t_* and whose radius is $|t_0-t_*|$.

Therefore, the appearance in $\mathbb{C}$ of a natural boundary is purely fictitious. It is an effect of projecting our Riemann surface onto the complex plane. Under this projection isolated singularities become dense on a circle.

Now, suppose that $\eta \neq 0$. Then, the general situation remains the same but the geometrical details change. In the complex plane $\mathbb{C}$ the poles t_n lie on a spiral winding about t_*. This spiral, of course, is a fractal of Hausdorff dimension zero. On the Riemann surface the poles t_n are isolated singularities lying on a spiral emanating from t_*. Each sheet contains only finitely many such singularities, and no fractal is produced. In fact, the number of poles t_n on each sheet is $[1/\delta]$. This result is obviously true for the previous case $\eta = 0$.

4-C. In these paragraphs we shall prove that the singularities produced by our canonical example (3.2) are isolated poles in the appropriate Riemann surface accumulating at no more than two points. In 4-E below we shall prove that, in a certain special case, the image under projection into the complex plane of these poles gives the appearance of being a natural boundary. With this in mind we can call that image a <u>pseudo-natural boundary</u>.

To begin, each of the functions $A_j (t - t_*)^{\alpha_j}$ in (3.9) is analytic everywhere except at the points $t = t_*$ and $t = \infty$. Therefore, the sum of finitely many such functions is analytic everywhere on the Riemann surface except at $t = t_*$ and $t = \infty$. In particular the zeroes of

$$x_1 (t)^{-1} = \sum_{j=1}^{n} A_j (t - t_*)^{\alpha_j} \tag{4.6}$$

necessarily remain isolated except at t_* or ∞. Otherwise,
they would produce essential singularities at their points
of accumulation and the sum would not be analytic.

As a consequence of all this, the function

$$x_1(t) = \left\{ \sum_{j=1}^{n} A_j (t - t_*)^{\alpha_j} \right\}^{-1} \tag{4.7}$$

can only have isolated poles on its Riemann surface except for
the points $t = t^*$ and $t = \infty$, where poles can accumulate.

4-D. Here we consider the case

$$x_1(\tau) = \{A_1 \tau^{\alpha_1} + A_2 \tau^{\alpha_2} + A_3 \tau^{\alpha_3}\}^{-1} \tag{4.8}$$

where we have set $\tau = t - t_*$. The moving poles are obtained
from the equation

$$A_1 + A_2 \tau^{\alpha_2 - \alpha_1} + A_3 \tau^{\alpha_3 - \alpha_1} = 0. \tag{4.9}$$

We choose $\alpha = \alpha_2 - \alpha_1$ and $\beta = \alpha_3 - \alpha_1$ to be real with
$0 < \alpha < 1$ and $0 < \beta < 1$, and we choose A_1, A_2, A_3 to be positive.

We assert that in the complex τ-plane, cut from $-\infty$ to 0,
there cannot be any poles. In other words, if $-\pi < \arg \tau < \pi$,
then τ is not a pole of (4.8).

To prove this assertion, we note that

$$\text{Im}\{A_1 + A_2 \tau^{\alpha} + A_3 \tau^{\beta}\} = A_2 |\tau|^{\alpha} \sin(\alpha \arg \tau)$$

$$+ A_3 |\tau|^{\beta} \sin(\beta \arg \tau) > 0. \tag{4.10}$$

whenever $0 < \arg \tau < \pi$. Therefore, there are no poles in the
half-plane $\text{Im } \tau > 0$. By complex conjugation there can be no
poles in the half plane $\text{Im } \tau < 0$. Finally, one can verify
that there are no poles on the positive real τ-axis.

4-E. Here we will construct the pseudo-natural boundary

indicated at the beginning of 4-C above. The occurrence of

this boundary will correspond to the presence of two rationally

independent Kowalevskaya exponents.

To do this, we specialize our considerations in 4-D to the

case in which $\alpha = 1$ and β is irrational. We can regard α and

β as our two rationally independent exponents. We also set

$A_3 = A_1 = 1$ and $\lambda = A_2$. Then, (4.9) becomes

$$\tau^\beta + \lambda\tau + 1 = 0. \tag{4.11}$$

If $\tau_0(\lambda)$ is a solution of (4.11), then

$$\tau_n(\lambda) = e^{\frac{2\pi i n}{\beta}} \tau_0\left(\lambda e^{\frac{2\pi i n}{\beta}}\right) \tag{4.12}$$

is also a solution of (4.11).

Suppose that

$$\tau_0(0) = e^{\frac{2\pi i}{\beta}} . \tag{4.13}$$

Then, one easily sees that $\tau_0(\lambda)$ is analytic about $\lambda = 0$.

Furthermore, one can prove that the projection of the poles

$\tau_n(\lambda)$ clusters on the image under a conformal mapping of the

circle $z = |\lambda|$ by the map

$$\tau = \frac{1}{\lambda} \phi_0(z) \tag{4.14}$$

where

$$\phi_0(z) = z\tau_0(z). \tag{4.15}$$

We now consider the solutions of (4.11) which are not

analytic at $\lambda = 0$. These behave like

$$\hat{\tau}_0(\lambda) \sim (-\lambda)^{\frac{-1}{1-\beta}}$$ (4.16)

near $\lambda = 0$. This new family of solutions can be obtained from a function $\bar{\tau}(z)$, analytic about $z = 0$, through the formula

$$\hat{\tau}_k(\lambda) = \lambda^{-\frac{1}{1-\beta}} e^{\frac{\pi i}{1-\beta}(2k+1)} \bar{\tau}\left\{ -\lambda^\beta e^{-\frac{2\pi i\beta}{1-\beta}(2k+1)} \right\}$$ (4.17)

$$(k = 0, \pm 1, \pm 2, \ldots).$$

The function $\bar{\tau}(z)$ itself is the solution of

$$-\bar{\tau} + \bar{\tau}^\beta + z = 0$$ (4.18)

analytic about $z = 0$ and equal to 1 at $z = 0$.

This second family of solutions produces poles which under projection cluster onto a closed curve obtainable as the image of the circle $z = |\lambda|$ by the map

$$\tau = -z^{-1} \bar{\tau}(z).$$ (4.19)

In the accompanying figure we exhibit the images under projection in the τ-plane of our two familes of poles.

ACKNOWLEDGMENT

One of us (D. Bessis) takes this opportunity to express his gratitude to Prof. M. F. Barnsley for a very fruitful discussion and to Dr. E. Vrscay for assistance in computational work. The authors of this paper are aware that their knowledge of the relevant literature may be incomplete. They take this opportunity to ask the interested reader to communicate appropriate additional references.

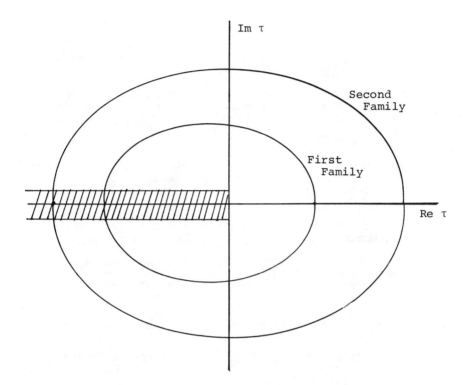

Fig. 1.   The projections of the isolated poles in the
Riemann surface onto the complex τ-plane cluster densely on
two closed curves when we have two rationally independent
Kowalevskaya exponents.

REFERENCES

[1]   U. Frisch and R. Morf, Phys. Rev. A. 23 (5), 1981.

[2]   M. Tabor and J. Weiss, Phys. Rev. A 24 (4), 1981.

[3]   Y. F. Chang, M. Tabor and J. Weiss, J. Math. Phys. 23
      (6), 1982.

[4]   J. Weiss, Analytic structure of the Henon-Heiles system,
      in Mathematical Methods in Hydrodynamics and Integrabi-
      lity in Related Dynamical Systems, ed. by M. Tabor and
      Y. Treve, AIP Conference Proc. 88, AIP, 1982.

[5]   I. C. Percival and J. M. Green, Hamiltonians Maps in
      the Complex Plane, Princeton Plasma Phys. Lab. Preprint,
      PPL 1744, Jan., 1981.

[6]   J. Weiss, M. Tabor, and G. Carnevole, J. Math. Phys. 24
      (3), 1983.

[7]   A. Ramani, B. Dorizzi, B. Grammaticos, and T. Bountis,
      J. Math. Phys. 25 (4), 1984.

[8]   T. Bountis and H. Segur, Logarithmic singularities and
      chaotic behavior in Hamiltonian systems, in Mathematical
      Methods in Hydrodynamics and Integrability in Related
      Dynamical Systems, ed. by M. Tabor and Y. Treve, AIP
      Conference Proc. 88, AIP, 1982.

[9]   T. Bountis, A singularity analysis of integrability and
      chaos in dynamical systems, in Singularities and
      Dynamical Systems, ed. by S. Pneumatikos, North Holland,
      1984.

[10] H. Yoshida, Necessary conditions for the existence of
      algebraic first integrals, Celestial Mechanics 31 (1983).

[11] H. Yoshida, Self-similar natural boundaries of non-
      integrable dynamical systems in the complex t-plane,
      Preprint, Dept. of Astronomy, Univ. of Tokyo, Tokyo,
      Japan.

[12] S. Kowalevskaya, Acta Math. Acad. Sci. Hungaria 14
      (1890), 81.

[13] Y. F. Chang, Program for ATOMCC, Claremont-McKenna
      College, Claremont, CA, U.S.A.

[14] O. Thual and U. Frisch, Natural boundaries in the
      Kuramoto model, in Workshop on Combustion Flames and
      Fires, Les Houches, France, 1984.

THE HENON MAPPING IN THE COMPLEX DOMAIN

John H. Hubbard

Department of Mathematics
Cornell University
Ithaca, New York

This paper reports on results of A. Douady, J. Hubbard and R. Oberste-Vorth.

1. INTRODUCTION

In this paper we will try to describe the behavior of the mapping $F: \mathbb{C}^2 \to \mathbb{C}^2$;

$$F: \begin{bmatrix} x \\ y \end{bmatrix} \to \begin{bmatrix} x^2 + c - ay \\ x \end{bmatrix}, \quad a \neq 0,$$

under iteration; these maps will be called the Henon family.

The mapping F has constant Jacobian a, and is invertible; in fact the inverse is given by

$$F^{-1}: \begin{bmatrix} x \\ y \end{bmatrix} \to \begin{bmatrix} y \\ \frac{1}{a}(y^2 + c - x) \end{bmatrix}$$

and is also polynomial of degree 2.

The mapping F is not quite as arbitrary as it might appear. Any polynomial map $F: \mathbb{C}^2 \to \mathbb{C}^2$ of degree 2 can be written $F = F_0 + F_1 + F_2$ with F_i homogeneous of degree i. In order for F to have constant Jacobian, it is necessary that F_2 have a one-dimensional kernel and a 1-dimensional image. Any automorphism $F: \mathbb{C}^2 \to \mathbb{C}^2$ for which the kernel and the image of F_2 are linearly independent can be conjugated to an element of the Henon family.

2. HISTORY AND MOTIVATION

a) Henon first considered the Henon family while trying to understand the Lorentz equations. He constructed a Poincaré section of the Lorentz equations, and then tried to find a simple mapping whose qualitative properties would be similar.

After some experimentation, he came up with an element of the Henon family which appeared to have a strange attractor similar to a section of the attractor found by Lorentz.

After much work, the theory is still fragmentary.

b) In 1925 (work completed by Bieberback in 1932) Fatou gave examples of injective analytic mappings

$$g: \mathbb{C}^2 \to \mathbb{C}^2,$$

whose images omit an open subset of $\mathbb{C}^2$.

The existence of such domains $U = g(\mathbb{C}^2)$ shows that many key results of complex analysis in one dimension fail in higher dimensions, for instance:

Montel: Any family $F_n: U \to \bar{\mathbb{C}}$ of meromorphic functions which omit three values is equicontinuous.

The following is one way (as far as I know the only way) of constructing such mappings g.

1. Find a mapping F of the Henon family with an attractive fixed point $\vec{x}_0$.

2. If the eigenvalues λ_i of $d_{\vec{x}_0} F$ satisfy $|\lambda_1| < |\lambda_2|$ and

$\lambda_1 \neq \lambda_2^n$ for all $n = 2,3,\ldots,$ then there exists

$g: (\mathbb{C}^2,0) \to (\mathbb{C}^2,\vec{x}_0)$ such that

$g(\vec{u}) = \vec{x}_0 + \vec{u} + o(|\vec{u}|)$, and

$F \circ g = g \circ d_{\vec{x}_0} F.$

This is a result of S. Sternberg ($\sim$1959); if $|\lambda_1| > |\lambda_2|^2$, you can show:

Proposition (Oberste-Vorth). The limit

$$g(\vec{u}) = \lim_{n \to \infty} F^{-n}((d_{x_0} F)^{\circ n}(\vec{u}) + \vec{x}_0)$$

exists for all $\vec{u} \in \mathbb{C}^2$, and defines a Fatou-Bieberbach mapping.

If $|\lambda_1|$ and $|\lambda_2|$ are farther apart, it is harder to construct g.

Such mappings have largely been viewed as pathological. Calabi asked me what the image looked like. The results of this paper grew out of his question.

3. THE RELATION WITH THE THEORY OF POLYNOMIALS

In the study of iteration of polynomials of one variable, extending to complex values of the variable has been very useful, even if the original polynomials was real. We hope the same thing will happen in this case, essentially for the same reason.

There is essentially nothing you can say about real polynomials which is independent of the coefficients, largely because virtually all features independent of conjugation, such as periodic cycles, are liable to disappear as the

parameters are varied. In the complex domain, the behavior is
far more uniform.

Let $P(z) = z^d + a_{d-1}z^{d-1} + \ldots + a_0$ be a polynomial. The
most useful construction is the function $\phi_P(z): (\mathbb{C},\infty) \to (\mathbb{C},\infty)$
such that

$$\phi_P(P(z)) = (\phi_P(z))^d \text{ and}$$

$$\phi_P(z) = z + o(1) \text{ near } \infty .$$

The function $\phi_P(z)$ is constructed as follows

$$\phi_P(z) = \lim_{n\to\infty} (P^{\circ n}(z))^{1/d^n}$$

(This is a standard "scattering theory construction": go to
∞ by P, and return by the unperturbed map $P_0: z \to z^d$.) In
order to give a meaning to the $(1/d^n)$-power, write the limit
above as an infinite product

$$\phi_P(z) = z \cdot \frac{(P(z))^{1/d}}{z} \cdot \frac{P(^{\circ 2}(z))^{1/d^2}}{(P(z))^{1/d}} \cdot \ldots ,$$

and note that

$$\frac{(P^{\circ (k+1)}(z))^{1/d^{k+1}}}{P^{\circ k}(z))^{1/d^k}} = \left(1 + \frac{a_{d-1}(P^{\circ k}(z))^{d-1}+\ldots+a_0}{(P^{\circ k}(z))^d}\right)^{1/d^{k+1}} .$$

Since the denominator is larger than the numerator for large
z, we can define the root to be the principal branch. It is
easy to show that the infinite product converges.

If we don't want to worry about branches of roots, we
can define

$$h_P(z) = \lim_{n\to\infty} \frac{1}{d^n} \log_+ |P^{\circ n}(z)| ,$$

where $\log_+(x) = \sup(\log(x), 0)$.

It is quite easy to show that this limit exists and is continuous on all of $\mathbb{C}$, harmonic on $\mathbb{C} - K_p$, where

$$K_p = \{z \mid P^{\circ n}(z) \neq \infty\}.$$

In fact, h_p is the Green's function of K_p.

We will define functions ϕ and h analogous to these on $\mathbb{C}^2$.

4.   RATES OF ESCAPE FOR THE HENON FAMILY

Look at the formula for the Henon mappings. If x is reasonably large, and large with respect to y, then the predominant behavior is that the x-coordinate gets squared. That motivates the following proposition.

Proposition 1.   The limit

$$h_+[\begin{smallmatrix} x \\ y \end{smallmatrix}] = \lim_{n \to \infty} \frac{1}{2^n} \log_+ |F^{\circ n}[\begin{smallmatrix} x \\ y \end{smallmatrix}]_1|,$$

where $[\begin{smallmatrix} x \\ y \end{smallmatrix}]_1 = x$, exists and defines a continuous function on $\mathbb{C}^2$, harmonic on

$$U_+ = \{[\begin{smallmatrix} x \\ y \end{smallmatrix}] \mid h_+[\begin{smallmatrix} x \\ y \end{smallmatrix}] > 0\}.$$

We have

$$h_+(F[\begin{smallmatrix} x \\ y \end{smallmatrix}]) = 2h_+[\begin{smallmatrix} x \\ y \end{smallmatrix}].$$

The behavior of h_+ is partially described below.

Proposition 2.   The mapping

$$h_+ : U_+ \to \mathbb{R}_+$$

is a trivial fibration whose fibers are homeomorphic to the

complement of a solenoid in S^3.

The following sketch of a proof will explain what is meant by a solenoid.

Idea of Proof. We will first choose a region $V_+ \subset \mathbb{C}^2$ for which the points surely escape. One such choice is

$$V_+ = \{ [\begin{smallmatrix} x \\ y \end{smallmatrix}] \mid |y| < \alpha |x|^2, \ |x| > \beta \},$$

with $\alpha = \dfrac{1}{6|a|}$, and $\beta = \sup(\sqrt{2|c|}, \ 3\sqrt[3]{2}|a|, \ 3)$. Then $V_+ \subset U_+$, and $U_+ = \underset{n \geq 0}{\cup} f^{-n}(V_+)$. It is fairly clear that

$$h_+([\begin{smallmatrix} x \\ y \end{smallmatrix}]) = \log|x| + o(\log(|x|^2 + |y|^2)),$$

uniformly in V_+.

Let $U_+(s) = \{x \in U_+ \mid h_+(x) = s\}$, and $V_+(s) = U_+(s) \cap V_+$. Then $V_+(s)$ is for large s very nearly the set

$$\{ [\begin{smallmatrix} x \\ y \end{smallmatrix}] \mid |x| = e^s, \ |y| < \alpha |x|^2 \},$$

and as such is a solid torus. The key point is: How does $F(V_+(x)) \subset V_+(2s)$ look?

Answer: The map of f wraps $V_+(s)$ into $V_+(2s)$ by winding it around twice, as in the first picture on the following page.

Now we can think of

$$U_+ = V_+(s) \cup F^{-1}(V_+(2s) \cup F^{-2}(V_+(4s)) \cup \ldots,$$

where each of the terms in the increasing union is a solid torus winding twice around inside the next.

This is a little hard to imagine, but can be done if you recall that the outside, in S^3, of an unknotted solid torus is also a solid torus. So consider the second diagram:

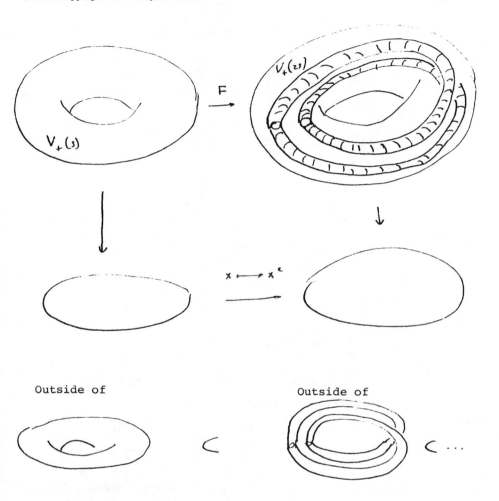

Outside of   Outside of

This realizes an increasing union as above, and the union is
the complement of a solenoid.   Q.E.D.

5. ANGLES OF ESCAPE

In the case of polynomials, there existed an analytic
function ϕ_P defined near ∞ such that

$$\log|\phi_P| = h_P.$$

Question: Can we define something analogous in $\mathbb{C}^2$?

Answer: No!   Still, what ought to be the fibers of ϕ do exist; however, they are dense in $U_+(s)$.   Their intersections with $V_+(s)$ are dense.

Proposition 3.   There exists an analytic function $\phi_+\colon V_+ \to \mathbb{C} - \bar{D}$ such that

$$\log|\phi_+| = h_+,$$

and

$$\phi_+(F(\vec{x})) = (\phi_+(\vec{x}))^2.$$

Idea of Proof.   Give a meaning to the roots in

$$\phi_+(\vec{x}) = \lim_{n\to\infty} ((F^{\circ n}(\vec{x}))_1)^{1/2^n}$$

by passing to an infinite product as in the case of polynomials, and show convergence.

You cannot extend ϕ_+ to all of U_+.   The following proposition describes what happens when you try.

Proposition 4.   A fiber of ϕ_+, as a closed Riemann surface in $V_+(s)$, can be continued to a Riemann surface isomorphic to $\mathbb{C}$ and dense in $U_+(s)$.   These Riemann surfaces foliate $U_+(s)$, and the mapping

$$\vec{x} \to \phi_+(\vec{x})/|\phi_+(\vec{x})|$$

induces a bijection of the set of leaves onto the (non-Hausdorff) group $\mathbb{R}/\mathbb{Z}$ $[\frac{1}{2}]$.

Idea of the Proof.   Consider again the picture

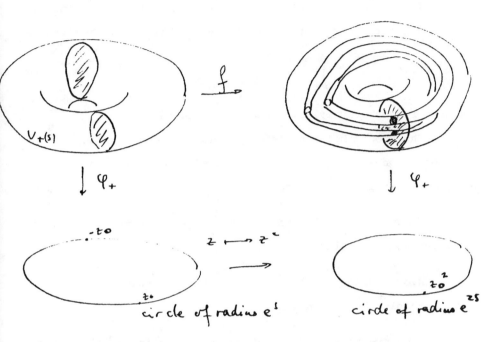

circle of radius e^s circle of radius e^{2s}

Since F is analytic, we see that the two discs $\phi_+^{-1}(z_0)$ ∪
$\phi_+^{-1}(-z_0)$ are in $F^{-1}(V_+(2s))$ two subdiscs of the disc
$F^{-1}(\phi_+^{-1}(z_0^2))$. Continuing into $F^{-2}(V_+(4s)),\ldots$ we see that
$\phi_+^{-1}(z_0)$ can be continued to a Riemann surface with a stratification looking like the following drawing.

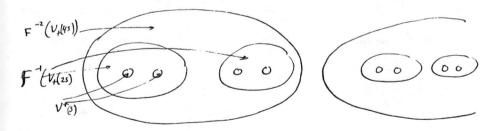

The shaded internal discs are the intersections of the
Riemann surface with the original $V_+(s)$; however, they are

$$\phi_+^{-1}\left(z_0 e^{\frac{2\pi i k}{2^m}}\right)$$

as m ranges over $1,2,\ldots$ and $k \in \{1,\ldots,m\}$.

The proof that the Riemann surfaces are isomorphic to $\mathbb{C}$
depends on showing that the moduli of appropriate annuli
grow; this is easy to show.   Q.E.D

Remark. The foliation is actually more natural than you
might expect. Since

Every real hyperplane of $\mathbb{C}^n$ contains a unique
complex hyperplane,

the tangent space to $U_+(s)$ at a point contains a unique complex direction. The leaves are simply the integral curves of
this field of directions. Of course, this could be said of
the level surfaces of any real valued function, but in
general such fields of directions have no integral curves.

6. A PROGRAM FOR DESCRIBING MAPPINGS IN THE HENON FAMILY

All the above could be said of F^{-1} as well as F, giving
rise to U_-, V_-, h_-.

Define $K_+ = \mathbb{C}^2 - U_+$, $K_- = \mathbb{C}^2 - U_-$, $K = K_+ \cap K_-$.

We propose, as a strategy for studying a mapping F in the
Henon family, to try to understand how the fibers $U_+(s)$
collapse onto ∂K_+ as $s \to 0$.

As an example of such a description, we propose the following conjecture.

Let $\sigma: S^3 \to S^3$ be the map which is more or less implicit in the definition of a solenoid. More precisely define a sequence $T_i \subset S^3$, $i = \ldots, -1, 0, 1, \ldots$ of tori, each decomposing S^3 into two solid tori T_i^+ and T_i^-, with $T_i^+ \subset T_{i+1}^+$ and winding around twice. Suppose that $X^+ = \cap T_i^+$ and $X^- = \cap T_i^-$ are both solenoids, and define σ so that $\sigma(T_{i+1}^+) = T_i^+$.

On $\mathbb{R}^4$, parametrized in "spherical coordinates" by (r,s) with $r \in [0,\infty)$, $s \in S^3$, consider the map $g: (r,s) \to (r^2, \sigma(s))$.

Let $Y = (\mathbb{R}^4 - \text{cone}(X^-)) \cup B$, and note that $g(Y) \subset Y$ and Y is homeomorphic to $\mathbb{R}^4$.

Conjecture. If F has an attractive fixed point, then F is conjugate to $g: Y \to Y$ by a homeomorphism $\phi: \mathbb{C}^2 \to Y$.

This would have, among others, the consequence that there exists a Fatou-Bieberbach domain whose boundary is a topological manifold, and that there are infinitely many analytic embeddings of $\mathbb{C}$ into the boundary whose images are dense.

Computer pictures support the above conjecture.

DYNAMICAL COMPLEXITY OF MAPS

OF THE INTERVAL

Louis Block*

Department of Mathematics
University of Florida
Gainesville, FL

1. THE ŠARKOVSKII STRATIFICATION

This paper will describe some results (mostly of Coppel
and myself [6]) concerning the dynamics of maps of an interval.
We will let I denote a compact interval on the real line, and
C(I,I) denote the space of continuous maps from I to itself
with the topology of uniform convergence.

We will begin with a basic result due to Šarkovskii [21]
(see also [8], [20], and [22]). Let P_n denote those maps in
C(I,I) which have a periodic point of least period n. Then

$$P_3 \subset P_5 \subset P_7 \subset P_9 \subset \dots$$
$$\subset P_6 \subset P_{10} \subset P_{14} \subset P_{18} \subset \dots$$
$$\subset P_{12} \subset P_{20} \subset P_{28} \subset P_{36} \subset \dots$$
$$\dots$$
$$\subset \dots \subset P_8 \subset P_4 \subset P_2 \subset P_1 \ .$$

Also, if $P_i \subset P_j$ and $i \neq j$ then P_i is contained in the interior
of P_j [3].

---

*The author was J. H. Van Vleck visiting professor at
Wesleyan University during the preparation of this paper.

Chaotic Dynamics
and Fractals

Hart and I have obtained additional information in the
space $C^1(I,I)$ of continuously differentiable maps (with the
topology of uniform convergence of f and f').

Theorem. [9] If n is not a power of two then $P_n \cap C^1(I,I)$
is a closed subset of $C^1(I,I)$, and if n is a power of two then
the closure of $P_n \cap C^1(I,I)$ is contained in $P_{n/2}$.

This result yields information about families of maps.
Let f_s be a continuous one-parameter family of maps in $C^1(I,I)$
where the parameter s varies in some interval $a \leq s \leq b$.
Suppose, for example, that $f_b \in P_3$ and $f_a \notin P_2$. Then there
is some parameter s_5, with $a < s_5 < b$, such that $f_{s_5} \in P_5 - P_3$,
and there is a parameter s_7 with $a < s_7 < s_5$ such that $f_{s_7} \in$
$P_7 - P_5$, etc. We may visualize a stratification of $C^1(I,I)$
with strata $P_3, P_5 - P_3, P_7 - P_5, P_9 - P_7, \ldots, P_6 - (P_3 \cup P_5 \cup P_7 \cup \ldots)$,
$P_{10} - P_6, P_{14} - P_{10} \cdots \ldots, P_4 - P_8, P_2 - P_4, P_1 - P_2$. Any
continuous one parameter family of maps f_s in $C^1(I,I)$ which
passes from one of the strata to another must pass through all
of the strata in between.

2. TOPOLOGICAL ENTROPY

The Šarkovskii stratification provides one measure of the
dynamical complexity of a map of the interval. As we move
further along in the stratification (towards P_3) the dynamics
becomes (in some sense) more complicated.

A quantitative measure of dynamical complexity is given by
the topological entropy, defined by Adler, Konheim, and
McAndrew [1]. The (topological) entropy of a map f is a nonnegative real number, which we will denote by h(f). In

general, we may think of positive entropy as corresponding to
complicated dynamics.

Theorem. [13], [18] For a map f of the interval to itself,
$h(f) > 0$ if and only if f has a periodic point whose period is
not a power of two.

The "if" direction of this result was obtained by Bowen
and Franks [13]; the "only if" direction by Misiurewicz [18].
Maps of the interval with positive entropy, which are sometimes referred to as chaotic, may also be characterized by the
presence of horseshoes, homoclinic orbits, and subshifts of
finite type [5].

Sharp relations between topological entropy and the
Šarkovskii stratification can be given. For each positive
integer n, let M_n denote the infimum of all possible entropies
of a map of the interval having a periodic point of period n
(this minimal value may be attained as we shall see later).
Also, for each odd $q \geq 3$, let λ_q denote the unique positive
root of the polynomial $Lq(x) = x^q - 2x^{q-2} - 1$.

Theorem. [8] If $n = 2^s$ then $M_n = 0$, and if $n = q \cdot 2^s$,
where q is odd and $q > 1$, then $M_n = (\log \lambda_q)/2^s$.

3. TURBULENCE

We now define the notion of turbulence, which gives a
third measure of the dynamical complexity of a map of the
interval. We say $f \in C(I,I)$ is turbulent if there are closed
subintervals J and K of I with at most one common point such
that $J \cup K \subset f(J) \cap f(K)$. This terminology, due to Coppel

[15], was suggested by Theorem 1 of Lasota and Yorke [17].
Let T_n denote the set of $f \in C(I,I)$ such that f^n is turbulent.
Then Coppel [15] has shown that

$$T_1 \subset T_3 \subset T_5 \subset T_7 \subset T_9 \cdots$$
$$\subset T_2 \subset T_6 \subset T_{10} \subset T_{14} \subset T_{18} \cdots$$
$$\subset T_4 \subset T_{12} \subset T_{20} \subset T_{28} \subset T_{36} \cdots$$
$$\cdots$$

The following is also due to Coppel [15].

Theorem. [6], [15] $T_n = P_n$ if n is not a power of two,
and T_{2^s} is contained in the interior of $T_{3 \cdot 2^s}$. Also,
$T_n \cap C^1(I,I)$ is a closed subset of $C^1(I,I)$.

Thus, we obtain a turbulence stratification which refines
the Šarkovskii stratification for maps with positive entropy.
An equivalent refinement of the Šarkovskii stratification was
given by Blokh [12], using the notion of an L-scheme of
Šarkovskii [21].

We conclude this section with a recent result of Coven and
myself. A map of the interval is said to be transitive if
some point has a dense orbit.

Theorem. [7] If $f \in C(I,I)$ is transitive then f^2 is turbulent.

4. ENTROPY MINIMAL ORBITS

In this section we define certain types of periodic orbits
(of maps of the interval) and see how these relate to the
previous sections. First, we define the term simple orbit

for a periodic orbit of period n. If $n = 2^s$ the definition
may be given inductively as follows [4]. Any fixed point is
a simple orbit, and an orbit of f with period 2^s, where
$s > 0$, is simple if the left and right halves of the orbit
are simple orbits of f^2 with period 2^{s-1}. If $n > 1$ is odd
then an orbit of period n with midpoint x is simple if its
points have either the order (due to Stefan [22])

$$f^{n-1}(x) < f^{n-3}(x) < \ldots < f^2(x) < x < f(x) < \ldots < f^{n-2}(x)$$

or the reverse order. If $n = qm$, where $q > 1$ is odd and
$m = 2^s > 1$, then an orbit of period n is simple if it consists
of m blocks of q consecutive points, each block forming a
simple orbit of f^m (with period q), and the blocks themselves
being permuted by f like a simple orbit with period m.

Second, we define the term <u>strongly simple orbit</u>. Any
simple orbit of odd period or of period a power of two is
said to be strongly simple. A simple orbit of period $n = qm$,
where $q > 1$ is odd and $m = 2^s > 1$, is said to be strongly
simple if f maps the midpoint of each of the m blocks of q
consecutive points to another such midpoint, with one exception. This property implies that f maps each such block monotonically onto another, with one exception.

With this terminology a result of Coppel [14] can be
stated as follows. Similar results were obtained independently
by Ho [16] and Alseda, Llibre, and Serra [2].

<u>Theorem</u>. [14] If $f \in P_n$, but $f \notin P_k$ whenever $P_k \subset P_n$ and
$1 < k < n$, then every periodic orbit of f of period n is sim-
ple. In this case, if n is not of the form $3 \cdot 2^s$, where $s > 0$,
then every orbit of period n is strongly simple.

Coppel and I have sharpened this result using the turbulence stratification.

Theorem. [6] If $f \in T_n$, but $f \notin T_k$ whenever $T_k \subset T_n$ and
$1 < k < n$, then every orbit of f of period n is strongly simple.

Coppel and I have also strengthened a result of Hart and
myself [11] to obtain the following.

Theorem. [6] If $f \in C(I,I)$ has a periodic orbit of period
n then f has a strongly simple orbit of period n.

We remark that with the hypothesis of this theorem, f
need not have a unimodal orbit of period n.

We next define the entropy of a period orbit and the term
entropy minimal orbit. If $P = \{p_1, p_2, \ldots, p_n\}$ is a periodic
orbit of $f \in C(I,I)$, with $p_1 < p_2 < \ldots < p_n$, there is a unique
map L_p defined on the interval $[p_1, p_n]$ such that $L_p(p_i) =$
$f(p_i)$ for $i = 1, \ldots, n$ and L_p is linear on each interval
$[p_1, p_2], [p_2, p_3], \ldots, [p_{n-1}, p_n]$. We set $h(P) = h(L_p)$.

It follows from Misiurewicz and Szlenk [19] that $h(L_p)$
may be obtained as follows. Let $I_1 = [p_1, p_2]$, $I_2 = [p_2, p_3], \ldots,$
$I_{n-1} = [p_{n-1}, p_n]$, and let A be the (n-1) by (n-1) matrix with
entry 1 in row i and column j if $L_p(I_i) \supset I_j$ and entry 0 in
row i and column j otherwise. Then $h(L_p) = \log \lambda$, where λ is
the largest eigenvalue of A.

Note that h(P) may be defined equivalently as the infimum
of all entropies of maps $g \in C(I,I)$ such that g agrees with f
on P. We say P is entropy minimal if $h(P) = M_n$ where P has
period n (and M_n is defined in Section 2).

$\underline{\text{Theorem}}$. [6] Let P be a periodic orbit of $f \in C(I,I)$. P
is entropy minimal if and only if P is strongly simple.

It follows from this theorem and previous results that for
any continuous one parameter family of C^1 maps of the interval,
the first time an orbit of period n appears, it must be an
entropy minimal orbit.

5. HOMOCLINIC ORBITS

Homoclinic orbits, defined for maps of the interval in [5],
provide a fourth measure of dynamical complexity. Here, we
define two special types of homoclinic orbits (which may be
thought of as normal forms) due to Coppel [15].

Let H_n denote the set of $f \in C(I,I)$ for which there exist
points a, b, and c such that a is a periodic point of f of
period n, $f^n(b) = a$, $f^n(c) = b$, and either

$a < c < b,$

$f^n(x) > a$ for $a < x < b$, and

$x < f^n(x) < b$ for $a < x < c,$

or the same with the inequalities reversed. Let $H_n^{\#}$ denote the
set of maps $f \in C(I,I)$ for which there exist points $x_0, \ldots, x_4$
such that x_0 is a periodic point of f of period n, $f^n(x_k) = x_{k-1}$ for $k = 1,2,3,4$, and either

$x_1 < x_3 < x_0 < x_4 < x_2,$

$f^n(x) < x_0 < f^{2n}(x)$ for $x_0 < x < x_2$, and

$x < f^{2n}(x) < x_2$ for $x_0 < x < x_4,$

or the same with the inequalities reversed.

Theorem. [6] $T_1 = H_1$ and $T_{2^s} = H^{\#}_{2^{s-1}} \cup H_{2^s}$ for $s > 0$.
Also, $T_n = P_n = H_n$ if n is not a power of two.

This theorem, together with previous results, sharpens the
conclusions of Hart and myself [10] that:

1. If $f \in C(I,I)$ has positive entropy then f has homoclinic
    orbits of infinitely many periods.

2. For any continuous one parameter family of C^1 maps, having
    zero entropy for some parameter value and positive entropy
    for some larger parameter value, there are infinitely many
    parameters where homoclinic orbits of new periods are
    created.

Finally, we mention that if one restricts to quadratic-like
maps (i.e., unimodal maps with negative Schwarzian derivative)
then $T_{2^s} = H^{\#}_{2^{s-1}}$ [6], and thus, for quadratic-like maps, the
turbulence stratification is the same as the homoclinic
stratification obtained in [11].

ACKNOWLEDGEMENTS

Most of the results described herein are based on joint
research with Andrew Coppel [6] during a visit to Australian
National University this past summer. This paper also bene-
fitted from conversations with Ethan Coven at Wesleyan
University. It has been a great privilege for me to have had
the opportunity to work with both Andrew Coppel and Ethan Coven
during the past year.

REFERENCES

[1]   R. L. Adler, A. G. Konheim, and M. H. McAndrew,
      "Topological entropy," Trans. Amer. Math. Soc. 114 (1965),
      309-319.

[2]   L. Alseda, J. Llibre, R. Serra, "Minimal periodic orbits
      for continuous maps of the interval," Trans. Amer. Math.
      Soc. 286 (1984), 595-627.

[3]   L. Block, "Stability of periodic orbits in the theorem of
      Šarkovskii," Proc. Amer. Math. Soc. 81 (1981), 333-336.

[4]   ———, "Simple periodic orbits of mappings of the
      interval," Trans. Amer. Math. Soc. 254 (1979), 391-398.

[5]   ———, "Homoclinic points of mappings of the interval,"
      Proc. Amer. Math. Soc. 72 (1978), 576-580.

[6]   L. Block and W. A. Coppel, "Stratification of continuous
      maps of an interval," Preprint.

[7]   L. Block and E. M. Coven, "Topological conjugacy and
      transitivity for a class of piecewise monotone maps of
      the interval," Preprint.

[8]   L. Block, J. Guckenheimer, M. Misiurewicz, and L. S.
      Young, "Periodic points and topological entropy of one
      dimensional maps," Lecture Notes in Mathematics, vol.
      819, Springer-Verlag, Berlin and New York, 1980, pp.
      18-34.

[9]   L. Block and D. Hart, "The bifurcation of periodic orbits
      of one-dimensional maps," Ergod. Th. and Dynam. Sys. 2
      (1982), 125-129.

[10] ———, "The bifurcation of homoclinic orbits of maps of
      the interval," Ergod. Th. and Dynam. Sys. 2 (1982), 131-
      138.

[11] ———, "Stratification of the space of unimodal interval
      maps," Ergod. Th. and Dynam. Sys. 3 (1983), 533-539.

[12] A. M. Blokh, "An interpretation of a theorem of A. N.
      Šarkovskii," (Russian) Oscillation and stability of solu-
      tions of functional-differential equations, Inst. Mat.
      Akad. Nauk. Ukrain, SSR, Kiev, 1982, pp. 3-8.

[13] R. Bowen and J. Franks, "The periodic points of maps of
      the disk and the interval," Topology 15 (1976), 337-342.

[14] W. A. Coppel, "Šarkovskii-minimal orbits," Math. Proc.
      Camb. Phil. Soc. 93 (1983), 397-408.

[15] ———, "Continuous maps of an interval," Xeroxed notes.

[16]   C.-W. Ho, "On the structure of the minimum orbits of
        periodic points for maps of the real line," Preprint.

[17]   A. Lasota and J. A. Yorke, "On the existence of
        invariant measures for transformations with strictly
        turbulent trajectories," Bull. Acad. Polon. Sci. Ser.
        Sci. Math. Astronom. Phys. 25 (1977), 233-238.

[18]   M. Misiurewicz, "Horseshoes for mappings of the inter-
        val," Bull. Acad. Polon. Sci. Ser. Sci. Math. 27 (1979),
        167-168.

[19]   M. Misiurewicz and W. Szlenk, "Entropy of pieceiwse
        monotone mappings," Studia Math. 67 (1980), 45-63.

[20]   Z. Nitecki, "Topological dynamics on the interval,"
        Ergodic theory and dynamical systems II (College Park,
        MD, 1979-80), pp. 1-73, Progress in Math., 21, Birkhauser,
        Boston, 1982.

[21]   A. N. Šarkovskii, "Coexistence of cycles of a continuous
        mapping of the line into itself." (Russian) Ukrain Mat.
        Z. 16 (1964), 61-71.

[22]   P. Štefan, "A theorem of Šarkovskii on the existence of
        periodic orbits of continuous endomorphisms of the real
        line," Comm. Math. Phys. 54 (1977), 237-248.

A USE OF CELLULAR AUTOMATA TO

OBTAIN FAMILIES OF FRACTALS

Stephen J. Willson

Department of Mathematics
Iowa State University
Ames, Iowa

ABSTRACT

This paper is intended to be a simple exposition of how to use cellular automata to generate a family of related fractals. We shall discuss the nature of cellular automata and show how they may be used to obtain interesting limit sets. In the process we shall find some numbers in various ways, and then we shall try to relate the different numbers so obtained. Proofs will only be suggested, and the reader will be referred elsewhere for details. In this paper the emphasis will be on clarifying the general line of the argument.

1. A SHORT HISTORY OF CELLULAR AUTOMATA

The name "automaton" derives from the consideration of these objects by John von Neumann when he wished to study the inherent limitations on which machines could duplicate themselves. He originally tried to design self-replicating machines more realistically, but at a suggestion by Stanislav Ulam, he abstracted the idea into that of cellular automata.

Von Neumann's work was completed by Arthur Burks [16] and then
modified by E. F. Codd [5] and others, retaining this point of
view. Further essays using this approach may be found in
Burks [4].

The same objects, however, had also appeared in topological
dynamics without a characteristic name. In Gottschalk and
Hedlund [9] or Hedlund [12], for example, they appear as essentially all examples of continuous maps on certain infinite product spaces that commute with the natural shift maps.

Cellular automata achieved great popular attention in 1970
with the description of the "Game of Life" by Martin Gardner in
his Scientific American column [7]. The "Game of Life" is one
particular transition rule of a cellular automaton, invented
by John Horton Conway and noted for the apparent unpredicta-
bility of its behavior. Soon the "Game of Life" was being
simulated on a great many computers across the country and was
being subject to much popular informal scrutiny. Details on
the "Game of Life" may be found in Gardner [8] or in Berlekamp
et al. [3]. Suffice it to say that the "Game of Life" has been
found to constitute a universal computer, and consequently that
the problem of characterizing the specific behavior of all
cellular automata has thus been proved unsolvable.

More recent professional study of cellular automata has
been stimulated by the work of Stephen Wolfram [23]. He has
studied the statistical mechanics of cellular automata for use
in theoretical physics. He has also initiated some systematic
computer simulations of various cellular automata with the
object of classifying the general nature of their behavior [24].

Cellular automata in some form have been rediscovered and

utilized in other contexts as well. F. C. Hoppensteadt [14]
uses cellular automata to model the invasion of a population of
trees by an infection. Robert Axelrod [2] uses them to study
how strategies of cooperation may spread from neighbor to
neighbor. Other topics to which they have been applied include
the development of life [15], the organization of parallel-
processing computers [13], and the description of certain
reaction-diffusion systems [10], [11]. Further applications are
suggested in Aladyev's survey [1], and related work appears
in the entire first issue of Physica D, volume 10.

My own viewpoint has been geometric. For broad classes
of cellular automata one sees much regularity in the way the
patterns change and evolve. It is possible that certain cellular automata may be used to model the process of crystal
growth. My papers [18] and [19] study the mathematics of
cellular automata whose behavior mimics the growth of simple
polyhedral crystals; one sees evolution into objects with
planar faces from irregular seed crystals. More recently, my
papers [20] and [21] deal with cellular automata that may be
related to the growth of dendritic crystals -- crystals with
much feathery fine detail, like frost or snowflakes. Stephen
Wolfram and Norman Packard are making specific studies to
clarify the relationship.

2.   WHAT ARE CELLULAR AUTOMATA?

Cellular automata may be roughly described in terms of two
concepts -- configurations and transition rules. Roughly
speaking, a configuration is an assignment of a zero or a one to
each square of an infinite checkerboard. (More generally, one

may assign any member of a finite set to each square of an
n-dimensional checkerboard; but in this paper we shall restrict
our attention to the case of only zeroes and ones on a
2-dimensional checkerboard, for ease of exposition.) We might
use a symbol such as w to denote a configuration.

A transition rule is a map F which associates to any configuration w a new configuration Fw. It is helpful to think of
the configuration w as describing the state of a crystal. If
a particular square of the checkerboard contains a 1, think of
the corresponding site in the crystal lattice as being occupied;
while if the square contains a 0, think of the site as empty.
With this interpretation, a transition rule F corresponds to
the passage of time and gives the dynamics of crystal growth.
If w describes the crystal at one instant of time, then Fw
describes the crystal one unit of time later. Similarly,
$F^2w = F(Fw)$ describes the crystal two units of time after the
start of the clock. The fundamental problem is to describe the
state of the crystal F^kw as it is seen k units of time after
the start, for large k.

We shall deal only with transition rules which satisfy
certain niceness conditions. In particular, we shall assume
first that the rule is locally determined. This means that
each square in the checkerboard has a neighborhood consisting
of finitely many cells in a certain special geometric position
relative to the square in question. We assume that for any
configuration w and any square indexed by any (x,y), where x
and y are integers, whether Fw has a 1 at site (x,y) is determined entirely by the values in the configuration w at the
sites in the neighborhood of (x,y). Note that the squares in

the neighborhood need not be the adjacent squares. For
example, the neighborhood of the square indexed by (x,y), where
x and y are integers, might consist of all squares indexed by
(x + a, y + b), where a and b are allowed to vary independently
over all integers from -2 to 3 inclusive.

Secondly, we assume that our transition rule is <u>invariant</u>
<u>under translations</u>. This means that the same criteria are
used at any site (x,y) vis-a-vis its neighbors as any other
other site.

Third, we assume that ones appear only in the presence of
other ones. Thus, if the entire neighborhood of (x,y) consists only of 0's in the configuration w, then (x,y) will also
be assigned 0 in Fw. There is <u>"no spontaneous generation."</u>

Let us consider a particular example. Suppose G is the
transition rule described as follows: If w is any configuration and (x,y) is any square on the checkerboard, we shall say
that Gw assigns 1 to the square (x,y) in either of two situations. First Gw assigns 1 there if w already assigned 1 to
(x,y); more briefly, $Gw(x,y) = 1$ if $w(x,v) = 1$. Moreover,
$Gw(x,y) = 1$ if $w(x,y) = 0$ but $w(x,y-1) + w(x-1,y-1) +$
$w(x-2,y-1) = 1$ (mod 2). In this example the neighborhood of
(x,y) consists of the four squares (x,y), $(x,y - 1)$, $(x - 1, y - 1)$,
and $(x - 2, y - 1)$; these are the square itself together with the
three squares immediately below and then towards the left
either one or two steps.

In Figure 1 at the top we see an initial configuration w,
selected to be a single one in an infinite checkerboard of
zeros. At the bottom of Figure 1 we see $G^{15}w$. It is
immediately clear that the configurations will grow large with

```
00000000000000000000000000000000000000000000
00000000000000000000000000000000000000000000
00000000000000000000000000000000000000000000
00000000000000000000000000000000000000000000
00000000000000000000000000000000000000000000
00000000000000000000000000000000000000000000
00000000000000000000000000000000000000000000
00000000000000000000000000000000000000000000
00000000000000000000000000000000000000000000
00000000000000000000000000000000000000000000
00000000000000000000000000000000000000000000
00000000000000000000000000000000000000000000
00000000000000000000000000000000000000000000
00000000000000000000000000000000000000000000
00000000000000000000000000000000000000000000
00000000000000000000000000000000000000000000
00000000000000000000000000000000000000000000
00000001000000000000000000000000000000000000
00000000000000000000000000000000000000000000
00000000000000000000000000000000000000000000
```

```
000000000000000000000000000000000000000000000000
000000000000000000000000000000000000000000000000
00000011011011011011010110110110110110110 11000
000000101000101000101010001010001010000 0
00000011101110000011100000111011100000 00
00000010001000000010000001000100000000 00
00000011010110110101101101011000000000 00
00000010101000101010001010100000000000 00
00000011100001110000111000000000000000 00
00000010000001000000100000000000000000 00
00000011011011101101100000000000000000 00
00000010100010001010000000000000000000 00
00000011101110111000000000000000000000 00
00000010001000100000000000000000000000 00
00000011010110000000000000000000000000 00
00000010101000000000000000000000000000 00
00000011100000000000000000000000000000 00
00000010000000000000000000000000000000 00
000000000000000000000000000000000000000000000000
000000000000000000000000000000000000000000000000
```

Fig. 1 (top). A configuration w consisting of a single 1 in an infinite checkerboard of 0's.

Fig. 1 (bottom). The configuration $G^{15}w$ for w as in the top and for G the graph of a particular additive rule (mod 2) as described in the text. $Gw(0,0) = 1$ provided that either $w(0,0) = 1$ or else $w(0,0) = 0$ but $w(0,-1) + w(-1,-1) + w(-2,-1) = 1$ (mod 2).

time, and we may note the irregularity of the pattern. It
could be a difficult problem to predict exactly which sites
have 1's in $G^{1000000}w$.

Our example G has in fact a particular form: it is the
graph of an additive rule (mod 2). For a 2-dimensional
checkerboard, such rules F have the form that $Fw(x,y) = 1$ provided that either $w(x,y) = 1$, or else both $w(x,y) = 0$ and for
some choice of integers $n_1, n_2, n_3, \ldots$, (depending only on the
rule F in question) we have $w(x - n_1, y - 1) + w(x - n_2, y - 1) +$
$+ \ldots = 1$ (mod 2). Our principal theorems will be proved for
the cases where F is the graph of an additive rule.

3. RESCALING TO OBTAIN FRACTALS IN THE LIMIT

We noted in our example that the configurations $F^k w$ tend
to grow large as k gets large. Consequently, it is necessary
to rescale the configurations so that they may more readily
be compared. The growth being roughly linear with time, we
rescale $F^k w$ linearly with k as follows: For any k and any
initial configuration w we define Y_k to be the set of all
$(x,y)/k$ in the Euclidean plane such that $F^{k-1}w(x,y) = 1$. Thus
Y_k essentially locates the sites where a one appears in
$F^{k-1}w$, and then rescales by a factor k.

The sequence of sets Y_{2^j} for $j = 1,2,\ldots$ show considerable
regularity. In Figure 2 we show Y_{16} and Y_{32} for the case
where the initial configuration w consists of a single 1 in
square (0,0) and zeroes elsewhere, and where the transition
rule is the map G discussed in Section 1. Y_{16} is essentially
just a rescaled copy of $G^{15}w$ as seen in Figure 1. Note how the
effect of an increase in k by a factor of 2 seems to be an

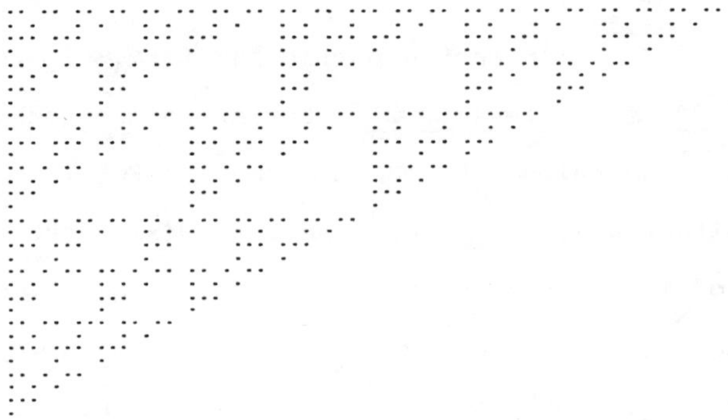

Fig. 2 (top). The set Y_{16} for the example in Figure 1.
This is a set showing the locations of the 1's in the bottom
of Figure 1. Rescaling makes the vertical distance be approximately 1 for each set Y_k.

Fig. 2 (bottom). The set Y_{32} for the example in Figure 1.
Note how it refines Y_{16}.

increase in the amount of fine detail in Y_k. In Figure 3
we show Y_{128} for the same example. The visible detail is still
finer than in Figure 2.

The tendency towards increase in finite detail is made precise in the following theorem:

Theorem 1. Suppose F is the graph of an additive rule
(mod 2) on the checkerboard. Suppose further that w is any
configuration with a finite positive number of ones, all of
which are on "one level" [in the sense that the sites (x,y)
possessing ones all have the same y value]. Then the finite
subsets $Y_2 k$ converge to some compact limit set Y.

The proof of Theorem 1 is in Willson [20]. The idea of the
proof is to exploit the algebraic identity that for any polynomial with mod 2 coefficients we have $p(x^2) = (p(x))^2$. The
fact that F is the graph of an additive rule mod 2 permits the
reduction of the problem to one dealing with such polynomials.
The notion of convergence is that of the Hausdorff metric.
(See Falconer [6].)   QED

In the example of the figures, the limit set Y will look
very much like Y_{128} in Figure 3, but will have additional
detail below our powers of easy resolution on the printed page.
By way of contrast, in Figure 4 , we see Y_{48}, which clearly is
not a good approximation to the same limit set Y. In fact, the
sets $Y_{3(2^k)}$ are also converging to a limit set W in the plane,
but this set W is different from the limit set Y. This situation is typical -- for a given transition rule F and an initial
configuration w there are many sets W such that certain subsequences $Y_{k(i)}$ converge to W.

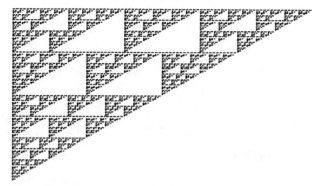

Fig. 3. The set Y_{128} for the example in Figure 1. Since the picture is near the limits of resolution, this is also our best approximation to the limit set Y.

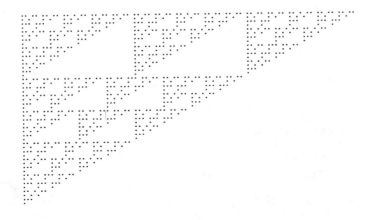

Fig. 4. The set Y_{48} for the example in Figure 1. Note that it does not give a refinement of Y_{32}. Indeed, the sets $Y_{(3)2^j}$ converge to a different fractal from that in Figure 3.

Theorem 2. Suppose F is a transition rule and w is any configuration with a finite positive number of 1's. Either the configurations $F^k w$ ultimately consist entirely of 0's, or else every sequence $Y_{k(i)}$ from the rescaled configurations has a convergent subsequence.

The idea of the proof is that all the sets Y_k are compact nonempty subsets of some fixed open bounded subset of the plane. The collection of all such subsets itself forms a compact metric space under the Hausdorff metric. See Falconer [6] for details of the topology.

The situation now stands as follows: If we are given a transition rule F and an initial configuration w, we obtain via Theorem 2 lots of different compact subsets Y of the plane, each a limit in the Hausdorff metric of some subsequence $Y_{j(i)}$ of rescaled configurations. Thus as far as we now have seen, the set Y may depend on the choice of w and the choice of the subsequence j(i), as well as on the mapping F. From Theorem 1 we see that the sets Y may be very interesting; indeed they often are fractals, as suggested in Figure 3. The remainder of the paper concerns how to make precise the relationship among the different limit sets Y and their dependence on the initial configuration w.

4. WAYS OF OBTAINING SOME NUMBERS FROM THE LIMIT SETS

Suppose w is a configuration with a finite positive number of ones and F is any transition rule. Form the rescaled configurations $Y_k = (F^{k-1} w)/k$, and suppose that we have a convergent subsequence $Y_{k(i)}$ converging to the limit set Y. In

this section we consider five numbers that may be obtained in
this situation.

1. Let D be the Hausdorff dimension of Y. (see Falconer
    [6] for the definition.)

2. Suppose that N(c) denotes the number of squares with
    side c meeting the set Y. (Since Y is compact, there
    are only finitely many such squares.)   Define
    the "capacity" or "upper Kolmogorov dimension" of Y
    to be D_+ = lim sup (log N(c))/(-log c) where the limit
    is taken as c approaches 0.

3. If N(c) is as in (2), define the "lower Kolmogorov
    dimension" D_- of Y to be lim inf (log N(c))/(-log c)
    as c goes to 0.

4. Suppose that M(k) denotes the number of points in Y_k
    or equivalently that M(k) is the number of ones in the
    configuration $F^{k-1}w$. Define the "lower growth rate
    dimension" D_1 to be lim inf (log M(k(i)))/(log k(i))
    as i goes to infinity.

5. If M(k) is as in (4), define the "upper growth rate
    dimension" D_2 to be lim sup (log M(k(i)))/(log k(i))
    as i goes to infinity.

The five numbers defined above all exist; and, for the
case of planar automata, all lie in the interval [0,2]. There
are certain obvious relations between them, such as $D \leq D_- \leq D_+$
and $D_1 \leq D_2$. In general these numbers may be distinct. Certainly from their definition we might expect each of them to
depend on the choice of the initial configuration w as well as
the choice of converging subsequence k(i). The main theorem
asserts that the five numbers all coincide and are independent

of the choice of k(i) and w.

Theorem 3. Suppose that w is a configuration possessing only a finite positive number of ones, all on one level. Suppose that F is the graph of an additive rule (mod 2). Pick any sequence k(i) such that the sequence $Y_{k(i)}$ converges, and call the limit Y. Then the following are true:

1. All five numbers defined above coincide, and we may call them by their common value D.

2. If we make a different choice of w satisfying the above conditions on w, the value of D remains unchanged.

3. If we make a different choice of subsequences k(i) such that $Y_{k(i)}$ converges (possibly to a different limit set), the value of D remains unchanged.

The theorem thus says that the various limit sets Y obtained in the manner of Section 3 are closely related. In general they are fractals, all of the same Hausdorff dimension. This dimension may be computed in several different ways. The growth rate dimension, which might be the easiest to estimate from a computer simulation, will necessarily give both the Hausdorff dimension and the capacity. This dimension is thus an invariant of the growth process given by F.

For an example to illustrate the independence of D from the choice of the initial configuration, consider Figure 5. This shows Y_{128} for the same example as in Figure 4, but with a different w. (Specifically, this w has ones in squares (0,0), (3,0), and (4,0) and zeroes elsewhere.) Note the close resemblance between the sets in Figures 4 and 5. In fact the

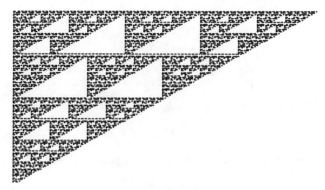

Fig. 5. The set Y_{128} for the same transition rule as in Figure 1 but with a different initial configuration w. The limiting sets Y will be the same as for the previous choice of w, but there are still noticable differences between this approximation and that in Figure 3.

limit sets Y in the two cases are identical. Furthermore, the common Hausdorff dimension may be readily computed using the growth rate dimension formulation, numbers (4) and (5). It is not too difficult to prove the recurrence relation

$$M(2^j) = 2M(2^{j-1}) + 4M(2^{k-2}).$$

This, together with the initial condition $M(2) = 4$ and $M(4) = 12$, lets us solve for $M(2^j)$ exactly:

$$M(2^j) = \frac{(5 + 3(5)^{1/2})(1+5^{1/2})^j}{10} + \frac{(5 + 3(5)^{1/2})(1 - 5^{1/2})^j}{10}$$

Thus we may conclude that $D_1 = \log(1 + 5^{1/2})/\log 2$. The transition rule G described above thus provides a way of creating a whole family of fractals, each of Hausdorff dimension $\log_2(1 + 5^{1/2})$

Proof. The proof of Theorem 3 is tedious, and details may be found in Willson [22]. First, the behavior of polynomials with coefficients mod 2 is utilized to obtain the fundamental inequality (where w has just a single square with 1)

$$M(2^{j+k}) \leq M(2^j)M(2^k).$$

An argument like that used to obtain the limit defining topological entropy [17] may then be applied to prove the existence of

$$\lim(\log_2 M(2^j))/j$$

as j goes to infinity. From this special case we may boot-strap to the existence and equality of the numbers D_1 and D_2; indeed we may show the existence of $\lim(\log M(k)/\log k)$ as k goes to infinity.

The next step of the proof is to compare the growth rate dimension $D_1 = D_2$ with the Kolmogorov dimensions D_+ and D_-. For this purpose an argument like that in Willson [21] suf-fices, so that we may conclude that these four numbers all coincide. But the Hausdorff dimension is always less than or equal to the capacity, providing one inequality. For the reverse inequality we use properties of polynomials mod 2 to show that the limit set Y will contain $M(2^j)$ appropriately rescaled copies of Y, one for each point in Y_{2^j} and originating at that point. Hence Y contains self-similar invariant sub-sets of Hausdorff dimension which may be shown to approach the capacity arbitrarily closely as j goes to infinity. Thus all five numbers coincide.

The proof that the Hausdorff dimension does not depend on the choice of w is another exercise in polynomial manipulation. To prove the independence from the choice of subsequence we can appeal to the existence of the limit $\lim(\log M(k))/\log k$,

which therefore gives the same growth rate dimension regardless
of the subsequence k(i).   QED

5. CONCLUSIONS AND DISCUSSION

   The theorems in this paper indicate that cellular automata
may be mathematically tractable objects for obtaining certain
fractals. Since it is illuminating to think of their dynamics
as describing the growth of a crystal, we should like to be
able to generalize the convergence results here to more general
transition rules F. Optimism suggests that if so the cellular
automata may be found useful in describing the growth of
dendritic crystals. In fact, preliminary studies suggest that
the results in Theorem 3 are true in a much more general context than that where F is the graph of an additive rule
(mod 2). The proofs we have given, however, rely on the algebraic properties of polynomials (mod p) where p is prime; thus
they usually do not generalize readily beyond that situation.
   The "Game of Life" provides examples which show that the
behavior may depend critically on the initial configuration w.
For some configurations w, F^kw dies out; for others it remains
bounded; and for still others it grows arbitrarily large.
Clearly then a generalization of Theorem 3 would very likely
have the dimensions dependent on the initial seed w. (Indeed,
the dimension will depend on the seed w even if F is the graph
of an additive rule (mod 2) provided that w is not on just one
level.) The dimensions might still, however, be independent
of the choice of subsequences if F satisfies a weak condition,
such as that no cell containing a one ever dies.

REFERENCES

[1]   V. Aladyev, Survey of research in the theory of
      homogeneous structures and their applications, Math.
      Biosci. 15 (1974), 121-154.

[2]   Robert Axelrod, "The Evolution of Cooperation," Basic
      Books, New York, 1984.

[3]   E. R. Berlekamp, J. H. Conway, and R. K. Guy, "Winning
      Ways for Your Mathematical Plays II. Games in Particu-
      lar," Academic Press, New York, 1982.

[4]   Arthur W. Burks, ed., "Essays on Cellular Automata,"
      University of Illinois Press, Urbana, 1970.

[5]   E. F. Codd, "Cellular Automata," Academic Press, New
      York, 1968.

[6]   K. J. Falconer, "The Geometry of Fractal Sets," Cambridge
      University Press, Cambridge, 1985.

[7]   M. Gardner, Mathematical games, Sci. Amer. 223 (October
      1970), 120-123; 224 (February 1971), 112-117.

[8]   M. Gardner, "Wheels, Life and Other Mathematical Amuse-
      ments," W. H. Freeman, New York, 1983.

[9]   W. H. Gottschalk and G. A. Hedlund, "Topological
      Dynamics," American Mathematical Society Colloquium
      Publications Vol. 36, Amer. Math. Soc., Providence, 1955.

[10] J. M. Greenberg and S. P. Hastings, Spatial patterns for
      discrete models of diffusion in excitable media, SIAM J.
      Appl. Math. 34 (1978), 515-523.

[11] J. M. Greenberg, B. D. Hassard, and S. P. Hastings,
      Pattern formation and periodic structures in systems
      modeled by reaction-diffusion equations, Bull. Amer.
      Math. Soc. 84 (1978), 1296-1327.

[12] G. A. Hedlund, Transformations commuting with the shift,
      in "Topological Dynamics," ed. by J. Auslander and
      W. H. Gottschalk, Benjamin, New York, 1968.

[13] W. D. Hillis, The connection machine: a computer
      architecture based on cellular automata, Physica 10D
      (1984), 213-228.

[14] F. C. Hoppensteadt, Mathematical aspects of population
      biology, in "Mathematics Today: Twelve Informal Essays"
      (L. A. Steen, ed.), Springer-Verlag, New York, 1978.

[15] S. A. Kauffman, Emergent properties in random complex
      automata, Physica 10D (1984), 145-156.

[16]   John von Neumann, "Theory of Self-Reproducing Automata,"
       edited and completed by Arthur W. Burks, University of
       Illinois Press, Urbana, 1966.

[17]   Peter Walters, "An Introduction to Ergodic Theory,"
       Springer-Verlag, New York, 1982, p. 87.

[18]   S. J. Willson, On convergence of configurations, Discrete
       Math. 23 (1978), 279-300.

[19]   S. J. Willson, Growth patterns of ordered cellular
       automata, J. Comput. System Sci. 22 (1981), 29-41.

[20]   S. J. Willson, Cellular automata can generate fractals,
       Discrete Applied Mathematics 8 (1984), 91-99.

[21]   S. J. Willson, Growth rates and fractional dimensions in
       cellular automata, Physica 10D (1984), 69-74.

[22]   S. J. Willson, The equality of fractional dimensions for
       certain cellular automata, Iowa State University pre-
       print, 1985.

[23]   S. Wolfram, Statistical mechanics of cellular automata,
       Reviews of Modern Physics 55 (1983), 601-644.

[24]   S. Wolfram, Universality and complexity in cellular
       automata, Physica 10D (1984), 1-35.

II. Julia Sets

EXPLODING JULIA SETS

Robert L. Devaney*

Department of Mathematics
Boston University
Boston, Massachusetts

ABSTRACT

Entire transcendental functions may exhibit what we call
explosions in their Julia sets as a parameter is varied. An
explosion is a sudden change from a nowhere dense Julia set to
one which is the entire complex plane. We illustrate this
phenomenon with an explosion in the complex exponential family,
and we describe the mechanism which produces explosions via
another example, the complex sine function.

1. INTRODUCTION

Our goal in this paper is to show how some simple local
bifurcations such as the saddle-node or period - doubling bifurcations may have global consequences when the dynamical systems
in question are complex analytic. We will show that the chaotic
sets for these types of systems may "explode" at such a bifurcation point.

In this paper we deal exclusively with entire transcendental functions which are <u>critically finite</u>. By critically

─────────
*Partially supported by the National Science Foundation.

141

finite, we mean functions which have only finitely many critical values and asymptotic values. We refer to [12] for a discussion of the role of these types of singular values in
dynamics as well as the basic definitions. Our methods very
definitely do not apply to polynomial maps.

We remark that the class of critical finite entire maps
includes such important maps as $\lambda \exp(z)$, $\lambda \sin(z)$, $\lambda \cos(z)$,
and $P(z)\exp(Q(z))$ where λ is a complex parameter and P and Q
are polynomials.

We will work with two simple but illustrative examples,
$E_\lambda(z) = \lambda \exp(z)$ and $S_\lambda(z) = \lambda \sin(z)$ where λ is a complex
parameter. We will describe a particular explosion which
occurs in the exponential family and will show that the
mechanism which leads to this explosion also applies to the
sine family.

Recall that the Julia set of a complex analytic map E is
the closure of the set of repelling periodic points of E. We
denote the Julia set of E by $J(E)$. There are other definitions of the Julia set in the literature (see [11], [1]), but
we will use this one. For maps like E_λ and S_λ, there is
another equivalent formulation of the Julia set, as described
in the following Proposition.

Proposition 1. Let $E = E_\lambda$ or S_λ. Then $J(E)$ is the closure
of $\{z \in C \mid E^n(z) \to \infty\}$.

Here E^n means the n-fold composition $E \circ \ldots \circ E$. We remark that
this proposition most definitely does not hold for polynomial
maps, as ∞ is always an attracting fixed point of such a map,
and its basin of attraction is not in the Julia set. It is

interesting that two quite different sets -- the set of repelling periodic points and the set of unbounded orbits -- have exactly the same closure. The proof of this Proposition is essentially contained in [9].

Let us recall some of the basic facts about Julia sets. We refer to [2] or [11] as background for this material.

1. $J(E)$ is a "strange repellor" which contains all of the chaotic dynamics of the map.

2. $J(E)$ is a closed, completely invariant, perfect set.

3. $J(E)$ does <u>not</u> contain any attracting periodic point or its basin of attraction.

4. Each attracting periodic orbit must attract at least one singular value (i.e., critical value or asymptotic value) of the map. Thus our assumption that E be critically finite implies that E has only finite many attracting periodic orbits.

5. The recent Non-Wandering Domain Theorem of Sullivan [14], as extended to our class of maps by Goldberg and Keen [12], states that, if all singular values tend to ∞ under iteration of E, then $J(E) = C$. We emphasize that there can be no "domains at infinity" (points which tend to ∞ but do not lie in $J(E)$ for the maps that we consider).

We define the stable set of $E, S(E)$ to be the complement of the Julia set. We think of S as containing all of the "tame" dynamics of the map while J contains the chaotic dynamics.

2. AN EXPLOSION IN THE EXPONENTIAL FAMILY

In this section we will describe an explosion that occurs in the family of maps $E_\lambda(z) = \lambda e^z$, where $\lambda > 0$ is a real parameter. The graphs of λe^x for x real assume two different forms depending upon whether $\lambda > 1/e$ or $\lambda < 1/e$. $\lambda = 1/e$ is the bifurcation point. See Figure 1.

Note that E_λ has no fixed points in $\mathbb{R}$ if $\lambda > 1/e$, and two fixed points, an attractor at q and a repellor at p, if $\lambda > 1/e$. This is the standard saddle-node bifurcation.

For the exponential map, there are no critical values and only one asymptotic value (at zero, the omitteḋ value). From Figure 1a we see that $E_\lambda^n(0) \to \infty$ when $\lambda > 1/e$, so it follows from the Sullivan Theorem that $J(E_\lambda) = C$ for $\lambda > 1/e$.

In case $\lambda < 1/e$, the Julia set is not the entire plane, for the attracting fixed point at $q \in \mathbb{R}$ must lie in $S(E_\lambda)$. What is $J(E_\lambda)$? We claim that the Julia set is dramatically different for these λ-values.

From Figure 1b we see that there is an interval between the attracting fixed point q and the repelling fixed point p such that, if $x_0 \in (q,p)$, then $\lambda e^{x_0} < x_0$. Now recall that E_λ

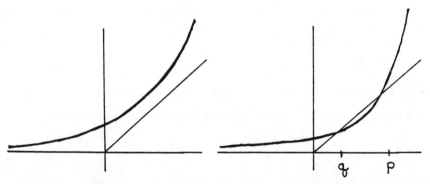

Figure 1a. $\lambda > 1/e$. Figure 1b. $\lambda < 1/e$.

maps vertical lines in C to circles centered at the origin.
Hence the vertical line $x = x_0$ is mapped to a circle of radius
$r = |\lambda| e^{x_0}$, and $r < x_0$. It follows that the entire half-plane
Re $z < x_0$ is mapped into the disk $|\lambda| < r$. Consequently, the
half-plane Re $z < x_0$ lies in the basin of attraction of q.
Thus, this half-plane lies in $S(E_\lambda)$.

    This, then, is an example of an explosion: when $\lambda > 1/e$,
$J(E_\lambda) = C$, but when $\lambda < 1/e$, $J(E_\lambda)$ is quite different, since it
misses the entire half-plane Re $z < x_0$.

    Let us briefly describe what the Julia set of E_λ is when
$\lambda < 1/e$. We claim that E_λ is a "Cantor bouquet." We will
describe $J(E_\lambda)$ by describing what it is not, i.e., by constructing the full stable set. Consider the vertical line
$x = p$. Clearly, if Re $z < p$, then the above arguments show
that $z \in S(E_\lambda)$. Now consider the preimage of the half-plane
Re $z < p$. This set consists of C minus a collection of tongues
as depicted in Figure 2a. These tongues each lie in a fundamental domain $(2n - 1)\pi < \text{Im}(z) < (2n + 1)$ where $n \in Z$ and are
mapped homeomorphically onto the right half plane H =
$\{z | \text{Re}(z) \geq p\}$. Hence E_λ expands any given tongue over all of
the others, and so the preimages of the tongues form an infinite
subcollection of tongues, as depicted in Figure 2b. Any point
which does not lie in this subcollection must eventually fall
into the basin of attraction of q and so is not in $J(E_\lambda)$.
Continuing in this fashion, one may check that the complement
of the basin of attraction of q is a Cantor set of curves.
This set is precisely the Julia set and the dynamics on this
set map may be described using symbolic dynamics. We again
refer to [9] for details.

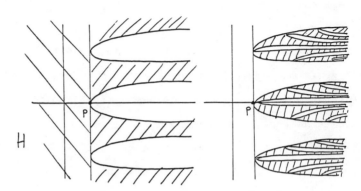

Figure 2a. $E_\lambda^{-1}(H)$. Figure 2b. $E_\lambda^{-1}(E_\lambda^{-1}(H))$.

Note that the entire interval $[p,\infty)$ must lie in $J(E_\lambda)$, for
any point $x \in (p,\infty)$ tends to ∞ under iteration of E_λ, and the
set of such points is contained in J by Proposition 1. This
half-line is an example of a "hair" in the Julia set which
will become important later.

At this point let us briefly indicate why the Julia set of
E_λ for $\lambda > 1/e$ is much "larger" than the case $\lambda > 1/e$. Let
B be any box in the left half plane of vertical height 2π and
arbitrary width. We will produce a point in B which maps onto
a repelling periodic point and hence lies in $J(E_\lambda)$.

This is best explained by a picture. Note that in Figure
3, E_λ maps B onto an annular region surrounding 0. Since
$E_\lambda^n(0) \to \infty$ it follows that E_λ successively maps points in this
region to the right. Indeed points in $E_\lambda(B) \cap \mathbb{R}$ tend to ∞
under iteration. Let us select a small disk D, contained in
$E_\lambda(B)$, and intersecting the real axis. E_λ maps D to the right
as shown in Figure 3. Moreover $|(E_\lambda^n)'(z)|$ must grow for
$z \in D$. Thus, eventually, E_λ^n must map D onto a set which over-

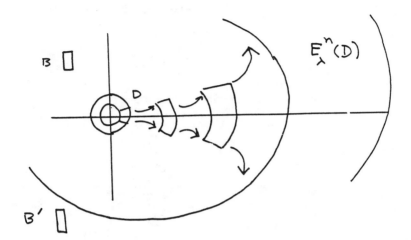

Figure 3.

laps the line $y = \pm\pi$. The next iteration of D is then a large
annular region in C. This annulus (or its next iterate) is
easily seen to overlap one of the translates of B by $2k\pi i$,
$k \in Z$. Call this box B'. Then we have that $E_\lambda(B) = E_\lambda(B')$,
and moreover, for some $N, E_\lambda E_\lambda^N(B')$ completely contains B' in its
interior. It follows that there must be a repelling periodic
point of period N inside B'.

Let us summarize this result:

Theorem A.   Let $E_\lambda(z) = \lambda \exp(z)$ for $\lambda > 0$. If $\lambda > 1/e$,
then $J(E_\lambda) = C$. If $\lambda < 1/e$, then $J(E_\lambda)$ is a nowhere dense
Cantor set of curves which form the boundary of a single basin
of attraction.

See [5] and [9] for further discussion. [3] contains
similar results.

Let us now briefly indicate the mechanism that has produced this explosion in the exponential family. See Figure 4.

Before

At the Bifurcation

After

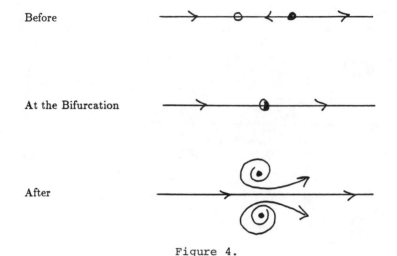

Figure 4.

Figure 4 shows that the saddle-node bifurcation for E_λ is somewhat different in the complex plane: the two fixed points q and p coalesce as λ increases through $1/e$, but they do not disappear. Instead, they enter the complex plane and become repellors. Since the derivatives at q and p must then be complex, we must have spiralling as shown in Figure 4. This allows 0 to escape through the tunnel between q and p and zip off to ∞. Alternatively, we view 0 as landing on the hair that had previously been attached to p and which consists of points which tend directly to ∞.

As a final remark we note that if λ becomes even slightly complex, the above argument breaks down. No longer need $E_\lambda^n(0)$ tend to ∞. Instead the iterates of 0 will tend to cycle around producing quite different effects and leading to the structural instability of E_λ. For details, we refer to [4].

3. AN EXPLOSION IN THE SINE FAMILY

Now let us turn to another type of bifurcation which produces an explosion, this time in the complex sine family.
Recall that

$$S_\lambda(z) = \lambda \sin z = \frac{\lambda}{2i} (e^{iz} - e^{-iz})$$

and that

$$S_\lambda(iy) = i\lambda \sinh(y).$$

Thus, as long as λ is real, S_λ preserves both the real and
imaginary axes in C. The graph of S_λ show that a bifurcation
occurs in this family when λ passes through 1. See Figure 5.

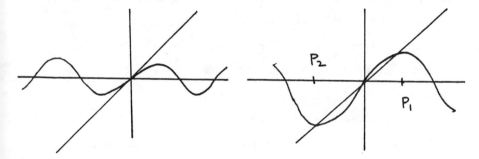

Figure 5a.   The graphs of $\lambda \sin x$ for $\lambda < 1$ and $\lambda > 1$.

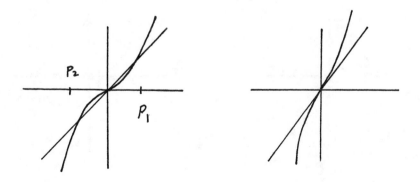

Figure 5b.   The graphs of $\lambda \sinh y$ for $\lambda < 1$ and $\lambda > 1$.

Note that, if $\lambda < 1$, 0 is an attracting fixed point and there are two repelling fixed points p_1 and p_2 on the imaginary axis for S_λ. Each of p_1 and p_2 come with hairs attached; that is, the intervals $[p_1, \infty)$ and $(-\infty, p_2]$ on the imaginary axis consist of points which tend to ∞ under iteration and which therefore lie in $J(S_\lambda)$.

When $\lambda > 1$, the situation is different. 0 is now a repelling fixed point, while p_1 and p_2 are now located on $\mathbb{R}$ and are attractors. All points on the imaginary axis except zero now tend to ∞ and thus $i\mathbb{R} \subset J(S_\lambda)$.

Thus, two things occur as λ crosses the bifurcation value $\lambda = 1$. First, 0 loses its stability and transfers it to p_1 and p_2. At the same time, p_1 and p_2 lose their hairs and transfer them to 0. Thus we think of this bifurcation as a "hair transplant." Note, for future reference, that both critical values λ and $-\lambda$ tend to 0 for $\lambda < 1$ and to p_1 and p_2 respectively if $1 < \lambda < \pi/2$. The phase portrait of S_λ is sketched in these two cases in Figure 6.

This bifurcation appears to be quite tame as we never lose basins of attraction and so there is no explosion. However,

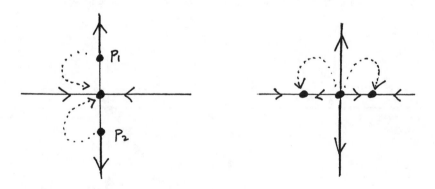

Figure 6.   The phase portrait of S_λ for $\lambda < 1$ and $\lambda > 1$.

let us look at the bifurcation diagram (i.e., the λ-plane) for
S_λ. Clearly, if $|\lambda| < 1$, S_λ has an attracting fixed point at
0 and it is easy to check that both critical values tend to
0 under iteration (use oddness of sin z). Hence S_λ has no
other attracting periodic points for $|\lambda| < 1$. We call this
region A. On the other hand there is a region B in the λ
plane for which there are two symmetric attracting fixed points
which we again denote by p_1 and p_2. Figure 7 displays the
relevant portion of the bifurcation diagram. Note that there
is a "wedge" W dividing A from B in the λ-plane. In W, each
of the points 0, p_1 and p_2 are repelling fixed points and one
may check that their derivatives are complex. Let us investigate the dynamics of S_λ for a λ-value in W. We claim that
there are parameter values arbitrary close to one in W for
which $J(S_\lambda) = C$.

To see this we consider the behavior of the hair which
extends to ∞ in the upper half-plane. One may easily show
that there is only one such hair in the strip $-\pi/2 < $ Re $z < \pi/2$

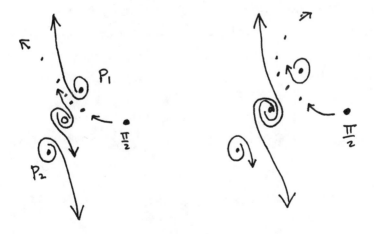

Figure 7. The dynamics of S_λ for λ near A and for λ
near B.

and Im $z \geq 0$. Hence there are only three possibilities: this
hair attaches to 0, to p_1, or to neither. For λ near the
region A, we have seen that the hair attaches to p_1, but for
λ near B, the hair attaches to 0. Let us then consider the
local dynamics to see what happens in between. Figure 7 shows
two possible scenarios.

Note that, since both 0 and p_1 are repelling, there is
again a tunnel in which the critical point $\pi/2$ may escape.
When the hair is attached to 0 (region B), $\pi/2$ necessarily
escapes to the right, while when the hair is attached to p_1,
just the opposite is true. Hence there must be an intermediate
case in which $\pi/2$ escapes directly toward ∞. To be precise,
there must be at least one λ-value in any curve cutting across
W in the λ-plane for which (1) $S_\lambda^n(\pi/2)$ lies in the strip
$-\pi/2 < \text{Re } z < \pi/2$ for all $n > 0$.   (2) $S_\lambda^n(\pi/2) \to \infty$ as $n \to \infty$.
By symmetry, $S_\lambda^n(-\pi/2)$ also tends to ∞, and so, for these
λ-values, $J(S_\lambda) = C$. We remark that these methods may be used
to show that there is a closed, connected set of λ values in
the strip $-\pi/2 < \text{Re}(z) < \pi/2$, $0 < \text{Im}(z)$ for which $J(S_\lambda) = C$.
Roughly speaking, this set is the frontier of the sets of
values for which $S_\lambda^n(\pi/2)$ escapes "to the right" or "to the
left."

   Theorem B. Let $S_\lambda(z) = \lambda \sin(z)$. In any neighborhood
U of 1 in the λ-plane, there exists λ_0 for which $J(S_{\lambda_0}) = C$.

4. CONCLUSION

The results in this paper tie in nicely with those of
Douady and Hubbard [7], [8], [10]. The external rays described
for quadratic polynomials in their papers find their counterparts in the hairs described above in the case of certain
entire maps. The hairs which we have described in the Julia
sets of entire functions are the exact analogues of the
external rays which lie in the basin of attraction of ∞ for a
polynomial. Note that there is a significant difference: in
the polynomial case, the rays <u>do not</u> lie in J, but for our
maps, by Proposition 1, they do.

A second point of contact with the work of Douady and
Hubbard involves the external rays to the Mandelbrot set
(i.e., the bifurcation set in the parameter space). Using
their techniques, one may show the set of parameter values in
the wedge W described in §3 is actually a curve, and so there
are "hairs" in the bifurcation set for $\lambda \sin(z)$ as well.
Note that these hairs consist of parameter values for which
$J(S_\lambda) = C$, which is quite different from the corresponding
situation for polynomials.

REFERENCES

[1] Baker, I. N., Repulsive Fixpoints of Entire Functions,
     Math. Z. 104 (1968), 252-256.

[2] Blanchard, P., Complex Analytic Dynamics on the Riemann
     Sphere, Bull. AMS 11 (1984), 85-141.

[3] Baker, I. N. and Rippon, P. J., Iteration of Exponential
     Functions, Ann. Acad. Sci. Fenn. Ser. I A Math. 9 (1984),
     49-77.

[4] Devaney, R., Structural Instability of Exp(z). To
     appear in Proc. AMS.

[5]   Devaney, R., Bursts into Chaos, Phys. Lett. <u>104</u> (1984),
       385-387.

[6]   Devaney, R., Julia Sets and Bifurcation Diagrams for
       Exponential Maps, Bull. AMS <u>11</u> (1984), 167-171.

[7]   Douady, A. and Hubbard, J., Itération des Polynomes
       Quadratiques Complexes, CRAS <u>294</u> (Janvier 1982).

[8]   Douady, A. and Hubbard, J., Etude Dynamique des
       Polynomes Complexes, Publ. Math. D'Orsay (preprint).

[9]   Devaney, R. and Krych, M., Dynamics of Exp (z). Ergodic
       Theory and Dyn. Syst. <u>4</u> (1984), 35-52.

[10]   Douady, A., Algorithms for Computing Angles in the
       Mandelbrot Set, These Proceedings.

[11]   Fatou, P., Sur L'Itération des Fonctions Transcendantes
       Entières, Acta Math. <u>47</u> (1926), 337-370.

[12]   Goldberg, L. and Keen, L., A Finiteness Theorem for a
       Dynamical Class of Entire Functions, To appear.

[13]   Mandelbrot, B., The Fractal Geometry of Nature, Freeman,
       1982.

[14]   Sullivan, D., Quasi-Conformal Homeomorphisms and
       Dynamics I and III, Preprints.

ALGORITHMS FOR COMPUTING ANGLES

IN THE MANDELBROT SET

A. Douady
48 rue Monsieur le Prince
74006 Paris FRANCE

1. NOTATIONS

f_c: $z \mapsto z^2 + c$, $z \in \mathbb{C}$.

f^n = f iterated n times.

$K_c = \{z \mid f_c^n(z) \not\to \infty\}$ the filled-in Julia set.

$J_c = \partial K_c$ the Julia set

f_c has two fixed points $\alpha(c)$ and $\beta(c)$.

Convention: $\beta(c)$ is the most repulsive (the one on the right).

$M = \{c \mid K_c$ is connected$\} = \{c \mid 0 \in K_c\}$.

$\mathcal{D}_0 = \{c \mid 0$ is periodic for $f_c\}$ (centers of hyperbolic components).

$\mathcal{D}_2 = \{c \mid 0$ is strictly preperiodic$\}$ (Misiurewicz points).

$\mathcal{D}_1 = \{c \mid f_c$ has a rational neutral cycle$\}$ (roots of hyperbolic
  components).

155

2. POTENTIAL AND EXTERNAL ARGUMENTS

The potential G_c created by K_c is given by

$$G_c(z) = \lim \frac{1}{2^n} \text{Log}^+ |f_c^n(z)| = \begin{cases} 0 & \text{if } z \in K_c \\ \text{Log}|z| + \Sigma \frac{1}{2^{n+1}} \text{Log}\left|1 + \frac{c}{f^n(z)^2}\right| \end{cases}$$

$$\text{else}$$

The map $z \mapsto \phi_c(z) = \lim(f_c^n(z))^{1/2^n}$ is well defined for all $z \in \mathbb{C} - K_c$ if K_c is connected, only for $G_c(z) > G_c(0)$ if K_c is a Cantor set. The external argument with respect to K_c is $\text{Arg}_c(z) = \text{Arg}(\phi_c(z))$. (The unit for arguments is the whole turn, not the radian.) For $z \in J_c = \partial K_c$, one can define one value of $\text{Arg}_c(z)$ for each way of access to z in $\mathbb{C} - K_c$. The external ray $R(c, \theta) = \{z | \text{Arg}_c(z) = \theta\}$ is orthogonal to the equipotential lines.

The potential created by M is $G_M(c) = G_c(c)$. The conformal mapping $\phi_M : \mathbb{C} - M \to \mathbb{C} - \bar{D}$ is given by

$$\phi_M(c) = \phi_c(c)$$

This is the magic formula which allows one to get information in the parameter plane. This gives

$$\text{Arg}_M(c) = \text{Arg}_c(c) \quad \text{for } c \notin M.$$

The formula extends to the Misiurewicz points. For $c \in \mathcal{D}_0$, the situation is more subtle.

3. HOW TO COMPUTE $\text{Arg}_c(z)$ FOR $c \in \mathcal{D}_0 \cup \mathcal{D}_2$, z PREPERIODIC

Set $x_0 = 0$, $x_i = f_c^i(0)$, $z_1 = z$, $z_j = f^{j-1}(z)$, $\beta = \beta(c)$, $\beta' = -\beta(c)$. Join β, β', the points x_i and the points z_j by arcs which remain in K_c. If such an arc has to cross a

component of U of $\overset{\circ}{K}_c$, let it go straight to the center and
then to the exit (the center of U is the point of U which is
in the inverse orbit of 0, "straight" is relative to the
Poincaré metric of U). You obtain a finite tree. Drawing
this tree requires some understanding of the set K_c and its
dynamics, but then it is enough to compute the required angle.
Choose an access ζ to z, and let ζ^j be the corresponding
access to z_j. The __spine__ of K_c is the arc from β to β'. Mark
0 each time ζ^j is above the spine and 1 each time it is below.
You obtain the expansion in base 2 of the external argument
θ of z by ζ. This simply comes from the two following facts:
a) $0 < \theta < 1/2$ if ζ is above the spine, $1/2 < \theta < 1$ if it
    is below;
b) f_c doubles the external arguments with respect to K_c, as
    well as the potential, since ϕ_c conjugates f_c to $z \mapsto z^2$.
Note that if c and z are real, the tree reduces to the segment
$[\beta', \beta]$ of the real line, and the sequence of 0 and 1 obtained
is just the __kneading sequence__ studied by Milnor and Thurston
(except for convention: they use 1 and -1). This sequence
appears now as the binary expansion of a number which has a
geometrical interpretation.

4. EXTERNAL ARGUMENTS IN M

    If $c \in \mathcal{D}_2$, then the external arguments of c in M are the
external arguments of c in K_c. Their number is finite, and
they are rationals with even denominators.

    A point $c_0 \in \mathcal{D}_0$ is in the interior of M, thus has no
external arguments. But the corresponding point $c_1 \in \mathcal{D}_1$
(the root of the hyperbolic component whose center is c_0) has

2 arguments θ_+ and θ_-, which are rational with odd denominators.
They can be obtained as follows: Let U_0 be the component of
the interior of K_{c_0} containing 0 and $U_i = f_c^i(U_0)$, so that U_1
contains c_0. On the boundary of U_0, there is a periodic point
α_0 whose period divides k = period of 0. Then θ_+ and θ_- are
the external arguments of $\alpha_1 = f_{c_0}(\alpha_0)$ corresponding to
accesses adjacent to U_1.

5.   USE OF EXTERNAL ARGUMENTS

We write $\theta \sim_c \theta'$ if the external rays $R(c,\theta)$ and $R(c,\theta')$
land in the same point of K_c, i.e., if θ and θ' are two
external arguments of one point in J_c. For $c \in \mathcal{D}_0 \cup \mathcal{D}_2$, the
classes having 3 elements or more is made of rational points,
each class with 2 elements is limit of classes made of
rational points. Knowing this equivalence relation, one can
describe K_c as follows: start from a closed disc, and pinch
it so as to identify the points of argument θ and θ' each time
you have $\theta \sim_c \theta'$. You end up with a space homeomorphic to K_c.
A similar description can be given for M. It will be valid if
we know that M is locally connected (a fact which is highly
suggested by the many pictures of details of M that we have).

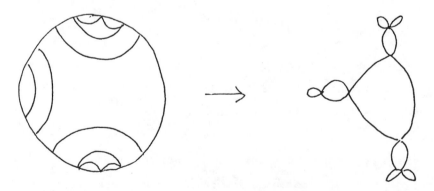

6. INTERNAL AND EXTERNAL ARGUMENTS IN ∂W_0

W_0 denotes the main component of M (the big cardioid).
Internal arguments in W_0 are defined using a conformal representation of W_0 onto the unit disc D. The internal argument
$\mathrm{Arg}_{W_0}(c)$ is just the argument of $f'_c(\alpha,c))$ (in fact the argument
of $\alpha(c))$. If a point $c \in \partial W_0$ has a rational internal argument
$t = p_0/q_0$ (irreducible form), a complment W_t of period q_0
is attached at W_0 at the point c. Thus c has 2 external
arguments

$$\theta_- = a_-/2^{q_0} - 1 \quad \text{and} \quad \theta_+ = a_+/2^{q_0} - 1.$$

Theorem 1.

$$\theta_-(t) = \sum_{s<t} 1/2^{q(s)} - 1 = \sum_{0<p/q<t} 1/2^q .$$

$\theta_+(t)$ = same with $\leq$ t instead of < t. (Here p(s)/q(s) is the
irreducible representation of the rational number s. In the
second sum, all representations are allowed.)

Proof. Clearly $\theta_+(t) - \theta_-(t) \geq 1/2^{q(t)} - 1$

$$\theta_-(t) \geq \sum_{s<t} 1/2^{q(s)} - 1.$$

$$1 - \theta_+(t) \geq \sum_{t<s<1} 1/2^{q(s)} - 1.$$

Lemma. $\sum_{s\in(0,1)\cap\mathbb{Q}} 1/2^{q(s)} - 1 = 1.$

Therefore the inequalities above are equalities, which
proves the theorem.

Proof of Lemma. Consider the integer points (q,p) with
$0 < p < q$, and provide each (q,p) with the weight $1/2^p$. Summing on horizontal lines gives total weight = 1. Summing on

rational lines through 0 gives total weight $= \Sigma_t \; 1/2^{q(t)} - 1$.

Corollary. The set of values of θ such that $R(M,\theta)$ lands on ∂W_0 has measure 0.

7. TUNING

Let W be a hyperbolic component of M, of period k, and c_0 the center of W. There is a copy M_W of , sitting in M, and in which W corresponds to che main cardioid W_0. This is particularly striking for a primitive component, and was observed by Mandelbrot in ~ 1980. More precisely, there is a continuous injection $\psi_W : M \to M$ such that $\psi_W(0) = c_0$, $\psi_W(W_0) = W$, $M_W = \psi_W(M)$, $\partial M_W \subset \partial M$. For $x \in M$, the point $\psi_W(x)$ will be called "c_0 tuned by x" and denoted $c_0 \perp x$ or $W \perp x$. The filled-in Julia set $K_{c_0 \perp x}$ can be obtained in taking K_{c_0} and replacing, for component U of $\overset{\circ}{K}_{c_0}$, the part $\bar{U}$ (which is homeomorphic to the closed disc $\bar{D}$) by a copy of K_x.

Theorem 2. Let θ_- and θ_+ be the two external arguments in M of the root c_1 of W, and let t be an external argument of x in M. Then to t there corresponds an external argument t' of $c_0 \perp x$ in M, which can be obtained by the following algorithm:
Expand θ_-, θ_+ and t in base 2:

$$\theta_- = .\;\overline{u_1^0 u_2^0 \ldots u_k^0} = .u_1^0 u_2^0 \ldots u_k^0 u_1^0 \ldots u_k^0 u_1^0 \ldots$$

$$\theta_+ = .\;\overline{u_1^1 u_2^1 \ldots u_k^1}$$

$$t = .s_1 s_2 \ldots s_n \ldots .$$

Then

K_{c_0}

K_x

$K_{c_0 \perp x}$

$$t' = .u_1^{s_1} \ldots u_k^{s_1}u_1^{s_2} \ldots u_k^{s_2}u_1^{s_3} \ldots$$

We denote this algorithm by $t' = (\theta_-, \theta_+) \perp t$. According to the principle: "You plough in the z-plane and harvest in the parameter plane," this theorem relies on Proposition 1 below. Let U_1 be the component of $\overset{\circ}{K}_{c_0}$ which contains $x_1 = x_0$, and α_1 the root of U_1 (the point on ∂U_1 which is repulsive periodic of period dividing k). Recall that θ_+ and θ_- are the external arguments of α_1 in K_{c_0} corresponding to the accesses adjacent to U_1.

Proposition 1. Let z be a point in ∂U_1 with internal argument t. Then z has an external argument t' in M given by $t' = (\theta_-, \theta_+) \perp t$.

Sketch of Proof of Proposition 1. Let α_1' be the point in U_1 opposite to α_1 (point of internal argument 1/2) and call the geodesic $[\alpha_1, \alpha_1']$ the spine of U_1. Let U_i be the connected component in $\overset{\circ}{K}_{c_0}$ containing $x_i = f_{c_0}^i(0)$, so that $U_i = f_{c_0}^{i-1}(U_1)$, and define the spine of U_i as the image of the spine of U_1. Recall that the spine of K_{c_0} is the arc $[\beta, \beta']$ in K_{c_0}.

Lemma. For each i, either U_i is off the spine of K_{c_0}, either the spine of U_i is the trace in $\bar{U}_i$ of the spine of K_{c_0}.

We don't prove this lemma here. Now, if the first digit s_1 of t is 0, z is on the same side of the spine of U_1 as the ray $R(\theta_-)$. Then $z_i = f_{c_0}^{i-1}(z)$ will be on the same side of the spine of U_i as $R(2^{i-1}\theta_-)$ (thus also on the same side of the spine of K_{c_0}) for $i = 1, \ldots, k$. Therefore the ith digit s_i' of t' is u_i^0. If $s_1 = 1$, then z follows θ_+, and $s_i' = u_i^1$ for

$i = 1,\ldots,k$. At time $k+1$, z_{k+1} is back in U_1, but with internal argument $2t$, so that s_1 is replaced by s_2 (the internal argument is preserved in the map from U_i to U_{i+1} for $i = 1,\ldots,k-1$, and doubled from $U_k = U_0$ to U_1). And we start for a second run, and so on...

Now let us come to the situation of Theorem 2. There is a homeomorphism ψ of K_x onto the part of $K_{c_0 \perp x}$ which corresponds to $\bar{U}_1$ in K_{c_0}. This homeomorphism conjugates f_x to $f^k_{c_0 \perp x}$.

<u>Lemma 2</u>. If $y \in J_x = \partial K_x$ has external argument t, then the external arguments of $\psi(y)$ in $K_{c_0 \perp x}$ are the same as the external arguments in K_{c_0} of the point $y' \in \partial U_1$ whose internal argument is t.

We don't prove this lemma here, but if you think of it, it is very natural. Theorem 2 now follows from results discussed in Section 4, applied to $y = x$ or to $y = $ the root of the component V_1 of $\overset{\circ}{K}_x$ containing x.

<u>Remark</u>. The copy M_W of M sits in M, but in some places, where M_W ends M goes on. Then there are points x in M with 1 external argument such that $c_0 \perp x$ has several external arguments. How is this compatible with the algorithm described in Theorem 2? Well, this algorithm is not univalent: It starts by expanding t in base 2, and numbers of the form $p/2^\ell$ have two expansions. Actually $c_0 \perp x$ may have more than 2 external arguments; the algorithm will give the two which correspond to accesses adjacent to M_W.

Example

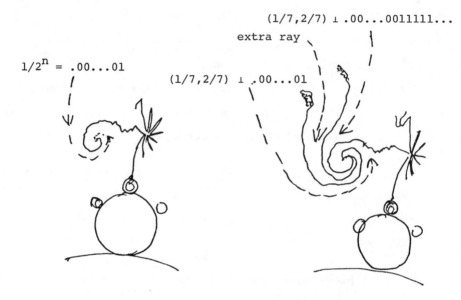

$(1/7, 2/7) \perp .00 \ldots 0011111 \ldots$

extra ray

$1/2^n = .00 \ldots 01$

$(1/7, 2/7) \perp .00 \ldots 01$

An elephant in M.

Its copy in W has two smokes coming out of its trunk if W is the rabbit component.

8. FEIGENBAUM POINT AND MORSE NUMBER

Let W_n be the n-th component of $\overset{\circ}{M}$ in the Myrberg-Feigenbaum cascade: W_1 is the disc $D(-1, 1/4)$ and we have $W_n = W_1 \perp W_1 \perp \ldots \perp W_1$. Denote $c_0(n)$ and $c_1(n)$ the center and the root of W_n. According to Feigenbaum theory, $c_0(n)$ converges to $c_\infty = -1.401 \ldots$ exponentially with ratio $1/4.66 \ldots$. The external arguments $\theta_-(n)$ and $\theta_+(n)$ of $c_1(n)$ are obtained by Theorem 2, starting from

$$\theta_-(1) = 1/3 = \overline{.01} \quad \text{and}$$

$$\theta_+(1) = 2/3 = \overline{.10} \ .$$

One gets:

$$\theta_-(2) = .\ \overline{0110}$$

$$\theta_+(2) = .\ \overline{1001}$$

$$\theta_-(3) = .\ \overline{01101001}$$

$$\theta_+(3) = .\ \overline{10010110}$$

$$\theta_-(4) = .\ \overline{0110100110010110}$$

. . . .

The numbers $\theta_-(n)$ converge to a number $\theta_-(\infty)$ known as the
Morse number. (It has been proved to be transcendental by
Van der Pooten, an Australian number theorist in Bordeaux.
The idea is that a sequence of digits which represents an
element in $\mathbb{Z}/2\ [[T]]$ which is algebraic over $\mathbb{Z}/2\ (T)$ cannot
be the expansion in base 2 of an algebraic real number, except
if periodic.) Note that the convergence of $\theta_-(n)$ to $\theta_-(\infty)$ is
faster than any exponential convergence: the number of good
digits is doubled at each time. In our view, this is due to
the fact that, because of the growing of hairs, W_{n+1} is more
sheltered from Brownian dust by W_n and W_n is by W_{n-1}.

9. SPIRALING ANGLE

    Take $c \in \mathbb{C}$ (in M or in $\mathbb{C}$ - M), and let z_0 be a repulsive
periodic point for f_c, of period k and multiplier $\rho = (f_c^k)'(z_0)$.
There are external rays of K_c landing on z_0 (there may be a
finite number or an infinity of them, almost always a finite
number though). Take one of them R and let $\tilde{R}$ be an image of
R by a determination of $z \to Log(z - z_0)$. Recall that, with our

convention, $\text{Log } z = \text{Log}|z| + 2\pi i \text{ Arg } z$. When $z \to 0$ on R,
$w = \text{Log } z \to \infty$ in $\tilde{R}$ and $\text{Im } w/\text{Re } w = 2\pi\text{Arg}|z - z_0|/\text{Log}|z - z_0|$ has
a limit m that we call the <u>spiraling slope</u> of z_0. This slope
can be written $m = \dfrac{2\pi\sigma}{\text{Log}|\rho|}$, and σ will be the <u>spiraling number</u>
(or spiraling angle). We have $\sigma = \text{Arg } \rho - \omega$, where ω is the
rotation angle of the action of f^k on the set of external
rays landing at z_0. If there are ν such rays, then ω is of
the form p/ν with $p \in \mathbb{Z}$. Note that $\text{Arg } \rho$ and ω are but
angles, i.e., $\in \mathbb{T} = \mathbb{R}/\mathbb{Z}$, while σ is naturally in $\mathbb{R}$.

In order to compute ω and σ, let c vary in a simply connected domain $\wedge \subset \mathbb{C}$ and $z_0(c)$ vary accordingly, remaining
repulsive periodic of period k. We make the following observa-
tions:

1. ω is continuous in c, and is invariant under the Hubbard-
    Branner stretching procedure (here it means just sliding c
    along the external rays of M).

2. ω remains constant in $\wedge$ if the number of external rays
    landing in $z_0(c)$ is finite and constant.

3. σ tends to 0 when $|p| \to 1$, i.e., when z_0 turns indif-
    ferent periodic (note that m does not necessarily tend to
    0, and may well tend to ∞ so that you see the Julia set
    spiraling a lot).

If we take $c \in \mathbb{C} - W_0$ and $z_0(c) = \alpha(c)$, the least repulsive
fixed point of f_c, we obtain the following: is the internal
argument in W_0 of $\pi_{W_0}(c)$, which can be defined in the following
way if you admit that M is locally connected. If $c \in M$,
$\pi_{W_0}(c)$ is the point where an arc in M from c to 0 enters W_0.
If $c \notin M$, let $\pi_M(c)$ be the point where the external ray of M
through c lands in M. Then $\pi_{W_0}(c) = \pi_{W_0}(\pi_M(c))$. In fact,

one can modify this definition so that it does not depend on the local connectivity of M, and determines $\pi_{w_0}(c)$ unambiguously.

10. HOW TO DETERMINE $\pi_{w_0}(c)$ KNOWING 1 PAIR (t,t') SUCH THAT $t \sim_c t'$, $t \neq t'$

Here is the algorithm: expand t and t' in base 2:

$$t = .u_1 u_2 \ldots$$

$$t' = .u_1' u_2' \ldots$$

Set $\delta_i = u_i - u_i'$ mod 2. If the sequence δ_i ends in $1,0,0,0,\ldots$, i.e., $\delta_i = 0$ for $i \geq n$, $\delta_{n-i} = 1$, then $\theta = 2^n t = 2^n t'$ is the external argument of c in M, and the internal argument of $\pi_{w_0}(c)$ is the angle t_0 such that $\theta_-(t_0) \leq \theta \leq \theta_+(t_0)$ (the functions θ_- and θ_+ are defined in Section 6). If the sequence δ_i ends with $1111\ldots$, then $\pi_{w_0}(c) = -3/4$ (internal argument $1/2$). Else, look for 10 somewhere in the sequence, i.e., $\delta_{n-1} = 1$, $\delta_n = 0$. Then there is a t_0 such that $\theta = 2^n t$ and $\theta' = 2^n t'$ both belong to $[\theta_-(t_0), \theta_+(t_0)]$, and that is the internal argument of $\pi_{w_0}(c)$. Why? Let z be the point in K_c with t and t' as external arguments. The rays $R(2^{n-1}t)$ and $R(2^{n-1}t')$ are each on one side of the spine of K_c, therefore $f_c^{n-1}(z)$ belongs to this spine. Now the image of the spine $[\beta,\beta']_{K_c}$ is the arc $[\beta,c]_{K_c}$ from β to c in K_c, and because $R(2^n t)$ and $R(2^n t')$ are on the same side of the spine, necessarily $f_c^n(z) \in [\alpha(c),c]_{K_c}$. From this observation the result follows easily.

11. ACKNOWLEDGEMENTS

Sections 2 to 8 are a report on joint work with J. H. Hubbard.   Section 9 is a reflexion in common with Bessis, Geronimo, Moussa in Saclay, which might take a more precise form eventually.   Section 10 is due to Tan Lei, a Chinese student in Orsay.

THE PARAMETER SPACE FOR

COMPLEX CUBIC POLYNOMIALS

Bodil Branner

Mathematical Institute
The Technical University of Denmark
Lyngby, Denmark

1. INTRODUCTION

This paper reports on results of joint work with John H.
Hubbard.

Since any complex cubic polynomial can be conjugated to one
of form

$$P_{a,b}(z) = z^3 - 3a^2 z + b, \quad z \in \mathbb{C}, \ (a,b) \in \mathbb{C}^2$$

by an affine change of variables, the parameter space for cubic
polynomials is $\mathbb{C}^2$. The representation is chosen such that the
critical points in $\mathbb{C}$ are +a and -a. There are two simple critical points in $\mathbb{C}$ except when a = 0.

We will describe a rough decomposition of the parameter
space $\mathbb{C}^2$ due to different dynamical behavior under iteration.
But since most of the techniques work for polynomials of degree
d ≥ 2, we will start with some statements...

2. ABOUT POLYNOMIALS OF DEGREE d ≥ 2 IN GENERAL

The dynamical behavior under iteration of a rational map
is "dominated" by the behavior of the critical points.

Chaotic Dynamics
and Fractals

Therefore a rough partition of the parameter space can be based on the possible behaviors of the critical points.

For a polynomial ∞ is always a superattractive fixed point. Any complex polynomial of degree $d \geq 2$ can be written

$$P(z) = z^d + a_{d-2}z^{d-2} + \ldots + a_0$$

up to a change of variables.

The main tools for understanding iteration of a monic polynomial P are the ϕ_P-map and the h_P-map: <u>The</u> ϕ_P-<u>map</u> is defined in a neighborhood of ∞

$$\phi_P: (\mathbb{C},\infty) \to (\mathbb{C},\infty)$$

such that ϕ_P conjugates P to $z \mapsto z^d$

$$
\begin{array}{ccc}
(\mathbb{C},\infty) & \xrightarrow{\quad P \quad} & (\mathbb{C},\infty) \\
\phi_P \downarrow & & \phi_P \downarrow \\
(\mathbb{C},\infty) & \xrightarrow{\quad\quad} & (\mathbb{C},\infty) \\
& z \mapsto z^d &
\end{array}
\tag{1}
$$

and such that

$$\phi_P \text{ is tangent to the identity at } \infty. \tag{2}$$

The map ϕ_P is unique. The only other maps which satisfy (1) are multiplies of ϕ_P, $\nu\phi_P$, where ν is a $(d-1)$-th root of unity. <u>The</u> h_P-<u>map</u> $h_P: \mathbb{C} \to \mathbb{R}_+ \cup \{0\}$ is defined by

$$h_P(z) = \lim_{n\to\infty} d^{-n}\log_+|P^{\circ n}(z)|, \tag{3}$$

where

$$
\log_+(x) = \begin{cases} \log x & \text{if } x \geq 1 \\ 0 & \text{if } 0 \leq x \leq 1. \end{cases}
$$

The limit exists for all z, it is continuous on $\mathbb{C}$ and harmonic on $\mathbb{C} - K_P$, where

$$K_P = \{z \mid h_P(z) = 0\} = \{z \mid P^{\circ n}(z) \not\to \infty\}.$$

The map h_P measures the escape rate to ∞ or in other words the rate of attraction to ∞. It is easy to see that

$$h_P(P(z)) = dh_P(z).$$

Let Ω be the set of critical points in P in $\mathbb{C}$, and let H be the map

$$H(P) = \sup_{\omega \in \Omega} \{h_P(P(\omega))\} \tag{4}$$

measuring the supremum of attraction to ∞ for the set of critical values, $P(\Omega)$.

It can be shown, that ϕ_P can be extended to the set

$$U_P = \{z \mid h_P(z) > d^{-1}H(P)\}.$$

The map

$$\phi_P: U_P \to \mathbb{C} - \bar{D}_{\exp(H(P))^{1/d}}$$

is an analytic isomorphism. In U_P the connection between the ϕ_P-map and the h_P-map is the following

$$h_P(z) = \log|\phi_P(z)|.$$

Therefore the ϕ_P-map defines a canonical polar coordinate system on U_P (canonical, since ϕ_P is unique).

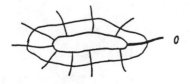

The "circles" are just the level curves for h_p, and radius ρ
correspond to rate of attraction $\log \rho$. The "radial lines,"
the rays, are the gradient lines for h_p. Every angle $\theta \in \mathbb{R}/\mathbb{Z}$
will be measured in turns and not in radians. In this polar
coordinate system on U_p the polynomial P is the mapping

$$\rho e^{2\pi i \theta} \mapsto \rho^d e^{2\pi i d \theta}.$$

3. DICHOTOMY FOR DYNAMICAL BEHAVIOR

The following theorem gives us a dichotomy for the dynamical behavior.

Theorem (Fatou, Julia). K_p _is connected if and only if_
$h_p(\omega) = 0$ _for all critical points_ $\omega \in \Omega$.

The parameter space for polynomials of degree d is $\mathbb{C}^{d-1}$.
It is decomposed into the Connectedness-Locus

$$L_d = \{\lambda \in \mathbb{C}^{d-1} | K_\lambda \text{ connected}\}$$

and the complement $\mathbb{C}^{d-1} - L_d$.

For _quadratic_ polynomials $P_c(z) = z^2 + c$ the Connectedness-
Locus equals the Mandelbrot set

$$M = \{c \in \mathbb{C} | K_c \text{ connected}\}.$$

For _cubic_ polynomials $P_{a,b}(z) = z^3 - 3a^2 z + b$ we shall call
the set

$$C = \{(a,b) \in \mathbb{C}^2 | K_{a,b} \text{ connected}\}$$

for the Cubic Locus. It is a double cover of the Connectedness-
Locus ramified along the intersection with the $\mathbb{C}$-line $a = 0$.

A. Douady and John H. Hubbard have proved the following theorem for quadratic polynomials.

Theorem. H: $\mathbb{C} - M \to \mathbb{R}_+$ is a trivial fibration and the fibers are homeomorphic to S^1.

Corollary. The Mandelbrot set is connected.

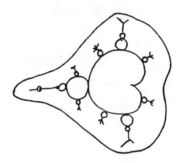

We have proved the similar theorem for cubic polynomials.

Theorem 1. H: $\mathbb{C}^2 - C \to \mathbb{R}_+$ is a trivial fibration and the fibers are homeomorphic to S^3.

Let S_r denote the fiber $H^{-1}(\log r)$, where $r > 1$.

Corollary. The Cubic Locus is cell-like, in particular connected.

This leads naturally to the following

Conjecture. H: $\mathbb{C}^{d-1} - L_d \to \mathbb{R}_+$ is a trival fibration and the fibers are homeomorphic to S^{2d-3}.

The technique used in the proof for the cubic polynomials is to stretch the complex structure and use Ahlfors-Bers theorem. The hard part is to show continuity of the trivializing map. There is one part in our continuity proof which would not work in degree > 3.

4. TRICHOTOMY FOR DYNAMICAL BEHAVIOR
    OF CUBIC POLYNOMIALS

Another important classical result by Fatou and Julia is
the following:

Let P be a polynomial of any degree d ≥ 2. If $h_P(\omega) > 0$
for every critical point $\omega \in \Omega$, then K_P is a Cantor set. The
dynamics on K_P is equivalent to a one-sided shift on d symbols.

For a quadratic polynomial $P_c(z) = z^2 + c$, the only critical point in $\mathbb{C}$ is 0. Therefore we have the following dichotomy
for the dynamical behavior:

1. 0 is attracted to ∞, in which case K_c is a Cantor set

2. 0 is not attracted to ∞, in which case K_c is connected.

For a cubic polynomial $P_{a,b}(z) = z^3 - 3a^2z + b$ we have the
following trichotomy for the dynamical behavior:

1. Both +a and -a are attracted to ∞, in which case $K_{a,b}$ is a
    Cantor set

2. Either +a or -a is attracted to ∞, but not both, in which
    case $K_{a,b}$ is disconnected; it usually has components not
    reduced to a point, but it might be a Cantor set.

3. Neither +a or -a is attracted to ∞, in which case $K_{a,b}$ is
    connected.

Our next question is therefore: how does the complement
of the Cubic Locus, $\mathbb{C}^2 - \mathcal{C}$, split up due to 1. and 2. in the
trichotomy? But before we answer that question we will give...

5. A TOPOLOGICAL DESCRIPTION OF S_r

The fiber S_r, for fixed $r > 1$, does naturally split into two parts

$$S_r = S_r^+ \cup S_r^- ,$$

where S_r^+ is the set of polynomials, where $+a$ is attracted to ∞ at least as fast as $-a$, and at rate $\log r$, and where S_r^- is the set where the roles of $+a$ and $-a$ are interchanged.

Restrict to S_r^+, such that $(a,b) \in S_r^+$ and $P = P_{a,b}$. Recall that ϕ_P is defined on

$$U_P = \{z \,|\, h_P(z) > 3^{-1}\log r\}.$$

Therefore $P(+a) \in U_P$ and we can read off the polar coordinates, say

$$\phi_P(P(+a)) = re^{2\pi i\alpha}.$$

Remark. The same is true for a quadratic polynomial P_c with $c \in \mathbb{C} - M$, and knowing r and α determines c uniquely.

We proceed for cubic polynomials by defining

$$\psi_r^+: \quad S_r^+ \to S^1$$

by

$$\psi_r^+(a,b) = r^{-1}\phi_P(P(+a)),$$

such that ψ_r^+ is the map which measures the angle for the critical value $P(+a)$.

Our next result is the following

Theorem 2. $\psi_r^+: S_r^+ \to S^1$ is a non-trivial fibration. The

fiber over $e^{2\pi i\alpha}$, $y_r^+(\alpha)$, is homeomorphic to a trefoil clover,
i.e., 3 closed discs with one point in common.

$y_r^+(\alpha)$:

The boundary of $y_r^+(\alpha)$ corresponds to the set of polynomials,
where +a and −a are attracted to ∞ at the same rate. The triple
point corresponds to the intersection of $y_r^+(\alpha)$ with the ℂ-line
$a = 0$.

If we interchange the roles of +a and −a, we get the simi-
lar result for $y_r^-(\alpha)$. The two trefoil clovers, $y_r^+(\alpha)$ and
$y_r^-(\alpha)$, are glued together along their boundaries.

Our third theorem describes the topological structure of
the triple $(S_r; S_r^+, S_r^-)$. Let S^3 be the standard unit sphere

$$S^3 = \{(z_1, z_2) \in \mathbb{C}^2 \mid |z_1|^2 + |z_2|^2 = 1\},$$

and decompose S^3 into the union of the two solid tori T_i where

$$T_i = \{(z_1, z_2) \in S^3 \mid |z_i|^2 \leq 1/2\}.$$

The intersection $T = T_1 \cap T_2$ is a torus, and the diagonal
$\Delta \subset T$ is an unknotted circle in S^3.

Theorem 3. The triple $(S_r; S_r^+, S_r^-)$ is homeomorphic to the
triple cover of $(S^3; T_1, T_2)$ ramified along Δ.

Since Δ is unknotted, this is a complete description up to
homeomorphism. (It is possible to visualize this construction
in $\mathbb{R}^3$, considered as the image of S^3 after a stereographic

projection.)

6. FINE STRUCTURE FOR $y_r^+(\alpha)$

The 3 leaves in the clover arise as follows:

Let $(a,b) \in$ int $y_r^+(\alpha)$, i.e., $P_{a,b}$ is a polynomial where +a is attracted to ∞ faster than -a and such that $\psi_r^+(a,b) = \alpha$ (-a might not be attracted at all).

In the <u>dynamical plane</u> the level curve for h_P at level $3^{-1} \log r$ is a figure eight, where +a is the singular point. <u>Two</u> rays meet at +a and these two rays correspond to 2 of the three thirds of α. Let $\alpha_1, \alpha_2, \alpha_3 \in \mathbb{R}/\mathbb{Z}$ be distinct and such that $3\alpha_1 = 3\alpha_2 = 3\alpha_3 = \alpha$. There are 3 possibilities in the dynamical plane.

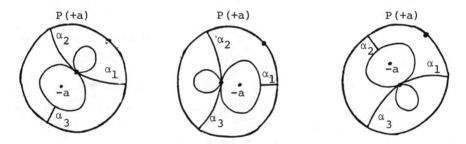

The leaves of $y_r^+(\alpha)$ in the <u>parameter space</u> correspond to

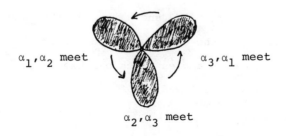

α_1, α_2 meet $\qquad$ α_3, α_1 meet

α_2, α_3 meet

After <u>one</u> turn of α, the 2 thirds α_1, α_2 have become α_2, α_3, after <u>another</u> turn α_3, α_1, and after a <u>third</u> α_1, α_2 again. This is the reason for the <u>triple cover</u>.

It is enough to understand <u>one</u> leaf, say the one where $\alpha = 0$ and $\alpha_1 = 1/3$ and $\alpha_2 = 2/3$. Every other leaf is homeo-morphic to this one, including all its interior structure. The technique is to <u>turn</u> the complex structure and the Ahlfors-Bers theorem.

The leaf contains infinitely many copies of the Mandelbrot set. Each copy correspond to a part, where we have condition 2. in the trichotomy satisfied. The Douady-Hubbard theory about polynomial-like mappings of degree 2 is essential. For each copy of M the center of the cardioid corresponds to $-a$ being periodic. The <u>main</u> Mandelbrot set is the one where $-a$ is a fixed point.

How the conditions 1. and 2. in the trichotomy partition the leaf can be described completely <u>combinatorially</u>. But the combinatorics involved is quite complicated. Each copy of the Mandelbrot set corresponds to a Cantor set construction in the set of angles for rays.

The amazing fact is that Cantor sets not only live in the dynamical planes but in the parameter space as well.

For $\alpha = 0$ and $\alpha_1 = 1/3$ and $\alpha_2 = 2/3$, the <u>main</u> Mandelbrot set corresponds to the classical middle-third Cantor set, i.e., the set of rays with angles which can be expressed in base 3 using only the digits 0 and 2.

The <u>main</u> Mandelbrot
set in the (1/3 , 2/3)-leaf
of $v_r^+(0)$.

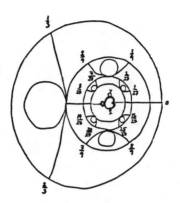

Remark. Change every 2-digit to a 1. Read off the number
in base 2. For every rational number p/q the ray converges to
the same boundary point of M as the ray $R(M , p/q)$ in the parameter plane $\mathbb{C}$ for quadratic polynomials.

7. FINAL REMARKS

Details will appear in two forthcoming papers: Iteration
of Complex Cubic Polynomials. Part 1, The Global Structure of
the Parameter Space; Part 2, Patterns and Parapatterns.

DISCONNECTED JULIA SETS

Paul Blanchard

Boston University
Department of Mathematics
Boston, Massachusetts

INTRODUCTION

The connectivity properties of the Julia set for a polynomial have an intimate relationship with the dynamical properties of the finite critical points. For example, if all
critical points iterate to infinity, then the Julia set J is
totally disconnected, and the polynomial p restricted to J is
topologically conjugate to the one-sided shift on d = deg(p)
symbols. On the other hand, if the orbits of all of the
finite critical points are bounded, then J is connected. In
this paper, we discuss other possibilities, and in particular,
we indicate how to construct symbolic codings for the components of the Julia set for a large class of cubic polynomials.
These cubics will have one critical point which iterates to
infinity and another whose orbit remains bounded. Using these
two orbits, we define a kneading sequence with two symbols, and
given certain kneading sequences, we show how to reconstruct
the dynamics of these cubics using the Douady-Hubbard [8]
theory of polynomial-like maps.

In Section 1, we establish our notation and summarize the
known results about symbolic codings and the Julia set with a

Chaotic Dynamics
and Fractals

particular emphasis on the quadratic case. Then, in Section 2, we describe our approach for cubics and its relationship to the Branner-Hubbard decomposition of the space of cubics. At the end of that section, we state a few important unresolved questions regarding the dynamics of cubics. For results that are implicitly used in this paper, the reader should consult the exposition [1], and the reader should also see the paper by Branner in this volume for a more elaborate discussion of the parameter space of cubics. In fact, we have made a serious effort to keep our notation consistent with that presentation.

1. NOTATION AND BACKGROUND MATERIAL

We consider polynomials $p(z)$ as functions of the Riemann sphere $\bar{C} = C \cup \{\infty\}$ and the associated discrete dynamical systems they generate. If $\deg(p) \geq 2$, the Fatou-Julia theory applies, and therefore, we have a disjoint, completely invariant decomposition

$$\bar{C} = J \cup N$$

where the Julia set J is the closure of the repelling periodic points and the domain N is the domain of normality for the family $\{p^n\}$ of iterates of the polynomial. The point at infinity plays a unique and important role. It is a super-attracting fixed point, and it belongs to a completely invariant component of N. That is, this component is invariant under both the map p and its inverse p^{-1}. In fact, this component also has another characterization in terms of its dynamical properties as the stable set of ∞:

$$W^s(\infty) = \{z \in \bar{C} \mid p^n(z) \to \infty \text{ as } n \to \infty\}.$$

Following Douady and Hubbard, we focus on the "filled-in" Julia set K of the polynomial defined by the equation

$$K = \bar{C} - W^s(\infty).$$

In this paper, we are mostly concerned with polynomials for which K is disconnected, and we describe the dynamics of the components of K using symbolic codings. Figures 1 and 2 illustrate two filled-in Julia sets for two different types of polynomials. The black regions are the filled-in Julia set and the shading of the stable manifold of infinity roughly corresponds to levels of the "rate of escape" map that is defined next.

We make frequent use of the "rate of escape to infinity" map $h: C \to R^+ \cup \{0\}$ defined by

$$h(z) = \lim_{k \to \infty} \frac{1}{d^k} \log_+ |p^k(z)| \qquad \text{where}$$

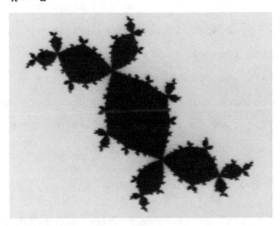

Fig. 1. The black region is the filled-in Julia set of the quadratic polynomial $z \mapsto z^2 + v$ where $v \approx -0.12256117 + 0.74486177i$. The value of v is chosen so that the origin (which is a critical point) is periodic of period three.

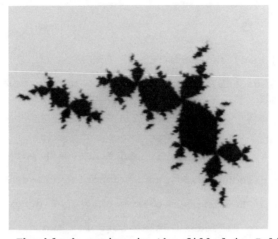

Fig. 2. The black region is the filled-in Julia set of the
cubic polynomial $z \mapsto z^3 - 3a^2z + b$ where $b = 0.8$ and
$a \approx -0.5769525 + 0.175i$. This kind of Julia set is the main
object of interest in this paper. Unlike the Julia set illustrated in Figure 1, one finite critical point escapes to
infinity and, therefore, the filled-in Julia set consists of
infinitely many components. We characterize these components
in Section 2.

$$\log_+(x) = \begin{cases} \log(x) & x \geq 1 \\ 0 & \text{otherwise.} \end{cases}$$

In fact, h is continuous on C and harmonic on $W^s(\infty)$. Note
$K = h^{-1}(0)$. Near infinity, we can give an alternate description using the conjugacy of p to the map $z \mapsto z^d$. Recall that
there exists a unique analytic homeomorphism $\psi\colon U_1 \to U_2$ where
U_1 and U_2 are open subsets containing infinity such that
$\psi(\infty) = \infty$, $\psi(p(z)) = [\psi(z)]^d$, and $D\psi_\infty = \text{Id}$. Using ψ, we obtain
another formula for h(z) as

$$h(z) = \log|\psi(z)| \quad \text{when } z \in U_1.$$

From this formula, it is easy to see that $h(p(z)) = d \cdot h(z)$.

Before discussing the unusual symbol spaces we use in the
second half of this paper, it is useful to elaborate on the

two extreme cases mentioned in the introduction. First of all,
we define the one-sided shift $\sigma|\Sigma_d$ on d symbols as the topo-
logical space

$$\Sigma_d = \prod_{k=0}^{\infty} \{1,2,\ldots,d\}$$

(where $\{1,\ldots,d\}$ is given the discrete topology and Σ_d is given
the associated product topology), and the shift map $\sigma: \Sigma_d \to \Sigma_d$
is the d-to-1 endomorphism defined by

$$[\sigma(\{s_i\})]_i = s_{i+1} \; .$$

Note that this map has d fixed points, many periodic points of
each period, points which are eventually periodic but not
periodic, and aperiodic points with dense orbits. Later, we
will find it easier to use symbols which are more mnemonic
than the numbers from 1 to d, but the symbol set will always
be equipped with the discrete topology.

Using $\sigma|\Sigma_d$, we can give a modern statement of the classi-
cal result concerning the two extremes mentioned in the introduction. Let C be the set of finite critical points of the
polynomial p(z).

Theorem 1. If $C \subset W^s(\infty)$, then $p|J$ is topologically con-
jugate to the map $\sigma|\Sigma_d$. On the other hand, if $C \subset K$, then J
is connected.

In the quadratic case, every polynomial is analytically
conjugate to one of the form

$$q_v(z) = z^2 + v.$$

In this form, $C = \{0\}$ for all q_v, and the first part of Theorem

1 applies if and only if

$$0 \mapsto v \mapsto v^2 + v \mapsto (v^2 + v)^2 + v \mapsto \ldots \to \infty.$$

When this happens, $h(0) > 0$, and using h, we can define the conjugacy $\phi: J \to \Sigma_2$ as follows. The level curve $L = h^{-1}(h(0))$ is a pinched circle which bounds two finite disks D_1 and D_2. Then

$$[\psi(z)]_k = \begin{cases} 1 <=> q^k(z) \in D_1 \\ 2 <=> q^k(z) \in D_2 . \end{cases}$$

This dichotomy (Theorem 1 applied to quadratics) motivates Douady and Hubbard's extensive analysis of the Mandelbrot set M. The set M is defined by

$$M = \{v \in C \,|\, J_{q_v} \text{ is connected}\},$$

and as Figure 3 indicates, it is remarkably complicated with an extremely interesting fractal structure.

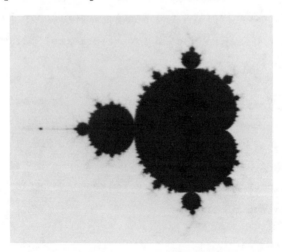

Fig. 3. The Mandelbrot set. See Mandelbrot [16, 17] for many more illustrations which indicate its complicated, yet regular structure. See also [9] and [10] for a detailed discussion of the dynamics of quadratics.

As we saw above, $\sigma | \Sigma_2$ can be used to study the dynamics of $q_v | J$ when $v \notin M$. An analogous question can be asked about the case where $v \in M$. This question was studied by Guckenheimer [12] and Jacobsen [14, 15] with the additional hypothesis that $p | J$ is expanding. Since their work is similar to what we describe in the next section, we give a brief statement of their results.

Theorem 2. (Jacobsen and Guckenheimer) If every critical point of a polynomial is attracted to some periodic sink (the point at infinity is allowed), then there exists a quotient with finite fibers of Σ_d which is topologically conjugate to $p | J$.

Consider two examples in the quadratic case. The first is the fundamental quadratic $z \mapsto z^2$. Recall that J is the unit circle. The quotient of Σ_2 in this case is generated by the identification $011\overline{1} \sim 100\overline{0}$. Extend this identification to all of Σ_2 in the minimal way such that there exists a well-defined quotient map $\sigma: (\Sigma/\sim) \to (\Sigma/\sim)$. In other words, $11\overline{1} \sim 00\overline{0}$, and $a_0 a_1 \ldots a_n 011\overline{1} \sim a_0 a_1 \ldots a_n 100\overline{0}$. The second example is the quadratic q_v whose Julia set is Douady's rabbit (see Figure 1 and recall that $\partial K = J$). In that case, we need one identification in addition to those of the first example to generate the quotient. It is $001\overline{001} \sim 010\overline{010} \sim 100\overline{100}$. One then adds the minimal set of identifications necessary to produce a well-defined quotient of the shift map.

2. SYMBOLIC CODINGS FOR CUBICS

In this section, we focus on the dynamics of cubics, and in particular on the intermediate case not covered in Theorem 1. Our approach is similar to that of Guckenheimer and Jacobsen in that we employ symbol sequences to study these maps, but it differs in that we associate a symbol for each component of K rather than for each point in K. The dynamics of the individual components is then determined using the Douady-Hubbard theory of polynomial-like maps. Although the coding in our approach is less precise than the Guckenheimer/Jacobsen coding, it has the advantage of applying to more general situations -- namely we do not need to make the assumption that $p|J$ is expanding.

For the remainder of the paper, we assume that p is a cubic and that the set of finite critical points consists of two distinct points c_1 and c_2 where $c_1 \in W^S(\infty)$ and $c_2 \in K$. First of all, consider the energy function and the structure of its level sets. As in the quadratic case, the level curve $L = h^{-1}(h(c_1))$ is a pinched curve (pinched at c_1) which bounds two finite disks -- denoted A and B. In fact, we fix our notation so that $c_2 \in B$, and therefore, $\deg(p|A) = 1$ and $\deg(p|B) = 2$. Given this decomposition, we define the A-B kneading sequence for p from the orbit of c_2 in K. This sequence $\{k_i\}$ is a sequence of the letters A and B by

$$k_i = \begin{cases} A & \text{if } p^i(c_2) \in A \\ B & \text{if } p^i(c_2) \in B \end{cases}$$

for $i = 0,1,2,\ldots$. In this section, we discuss a symbolic coding of the components of K for three distinct cases of $\{k_i\}$,

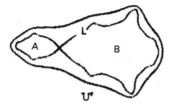

Fig. 4. The disk U' (for the cubic $z^3 - 1.47z + 0.8$)
defined by the rate of escape function h and its value for
$p(c_1)$. That is, $U' = h^{-1}\{[0,h(p(c_1))]\}$. The level curve L and
the two disks A and B that it bounds are all subsets of U'.

and then we state some open questions regarding the remaining

cases.

A key ingredient is the Douady-Hubbard theory of polynomial-
like maps [9]. Their idea is that often, in the dynamics of
high degree polynomials and even in transcendental functions,
one can find regions on which the dynamics is really determined
by a low degree polynomial.

Definition. Let U and U' be two simply-connected domains
in C such that U is a relatively compact subset of U'. If
f: U → U' is a proper, holomorphic map of degree d, then we
say that f is polynomial-like of degree d on U. Associated to
every polynomial-like map f: U → U', there exists a "filled-
in" Julia set K_f defined by

$$K_f = \{z \mid f^n(z) \in U \text{ for } n = 0,1,2,\ldots\}.$$

The basic theorem of Douady and Hubbard ([7] and [9])
which gives this notion its strength is the following result.

Theorem 3. (Douady and Hubbard) If f: U → U' is
polynomial-like of degree d, then there exists a polynomial q

of degree d such that $q|K_q$ is quasi-conformally conjugate to $f|K_f$.

This is precisely what is happening in the case at hand. We have two polynomial-like maps $f_1 = p|A$ and $f_2 = p|B$ such that $\deg(f_k) = k$. If the A-B kneading sequence of c_2 is BBB…, then $f_2|K_{f_2}$ has the dynamics of some quadratic q_v where $v \in M$, and if the kneading sequence is anything else, then $f_2|K_{f_2}$ is topologically conjugate to the one-sided shift on two symbols.

Before we state and prove our results, we should pause to relate this situation to the Branner-Hubbard decomposition of the parameter space of cubics described elsewhere in this volume. They discuss the dynamics of cubics in terms of the family

$$p_{a,b}(z) = z^3 - 3a^2z + b$$

where $(a,b) \in C^2$. The critical points are therefore $\pm a$. Let's suppose that $c_1 = +a$, and $c_2 = -a$. Then, according to their results, if we fix both the rate of escape of $+a$ to infinity (i.e., the value of $h(+a)$) and the angle of escape to infinity, then we are left with a two-dimensional subspace of cubics which is trefoil cloverleaf T. Moreover, the structure of parameter space inside of each leaf does not change as we move from leaf to leaf. Many cubics in T will also have $-a \in W^s(\infty)$, and consequently, Theorem 1 applies. However, it is interesting to relate the A-B kneading sequence to the structure inside T. Inside of each leaf of T, there exists an entire Mandelbrot set of cubics whose A-B kneading sequence is BBB... (see Figure 5) and another Mandelbrot set whose A-B kneading sequence is BABABA... (see Figures 5 and 6). In addition,

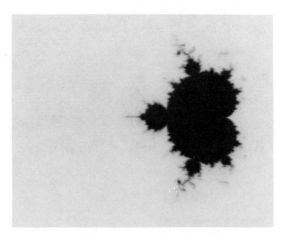

Fig. 5.   A slice of the cubic parameter space with b = 0.8.
The black regions indicate the areas where one critical point
escapes to infinity while the orbit of the other is bounded.
The largest Mandelbrot set corresponds to the kneading sequence
BBB... .

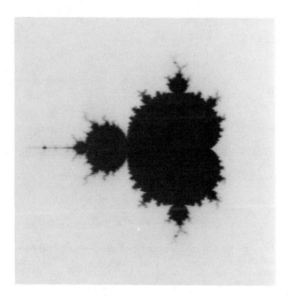

Fig. 6.   An enlargement of a small area along the horizontal axis of Figure 5.   The length of this side is about 3/1000
the length of the side in Figure 5.   The largest Mandelbrot set
corresponds to the kneading sequence BABABA... .

associated to the sequence BAAA..., there is exactly one cubic
in each leaf. These are the three cases for which we construct
a symbolic coding.

As one can immediately conclude from Theorem 3, the case
where the kneading sequence is BBB... differs widely from the
other possibilities. In this case K_{f_2} is a connected, filled-
in Julia set for a quadratic. Otherwise, K_{f_2} is a Cantor set.
We consider the BBB... case first.

Our symbol space describes the manner in which components
of K_p map, and we give a complete topological description of
$p|K_p$ if we know the dynamics of $f_2|K_2$. We use the symbol
space Σ' which is a σ-invariant subset of the one-sided shift
Σ_4 on the four symbols $\{1,2,3,B\}$. A sequence $s \in \Sigma'$ if and
only if

(1) $s_k = B \Rightarrow s_{k+1} = B$,

(2) $s_n = B$ and $s_{n-1} \neq B \Rightarrow s_{n-1} = 1$, and

(3) if $s_k \neq B$ for all k, then there exists a subsequence s_{n_i}
    such that $s_{n_i} = 1$ for all i.

The theorem states that there is a conjugacy between
$\sigma|\Sigma'$ and the space of components of K_p. As always, we use the
Hausdorff topology on the space of components. That is, two
components K_1 and K_2 are within ε of one another if every point
in K_1 is within ε of some point of K_2 and vice-versa. To state
the theorem, let $\tilde{p}$ denote the component-wise version of the map
p. In other words, if K_1 is a component of K_p, then $\tilde{p}(A)$ is
the component $p(A)$.

Theorem 4. Suppose p is a cubic polynomial whose Julia set
is disconnected but not totally disconnected. Using the notation of this section, suppose the A-B kneading sequence

associated to c_2 is BBB... . Then there exists a homeomorphism

$\phi: \Sigma' \to$ {components of K_p} such that $\phi \circ \sigma = \tilde{p} \circ \phi$. Moreover,

if $s \in \Sigma'$ is an element of Σ_3, then $\phi(s)$ is a point.

Remark. In all of the proofs of this section, we use
invariant sets of rays to infinity which are intimately connected with the conjugacy at infinity of $p(z)$ to the map
$z \mapsto z^3$. Basically, they are found in the following manner.
Near infinity, we have a polar coordinate system associated to
the cubic via the conjugacy ψ (as discussed in Section 1). A
ray of angle α is the inverse image under ψ of an angular ray
of angle $e^{2\pi i \alpha}$ from infinity in the standard coordinate system.
The only difficulty involving these rays comes when we try to
extend them so that they limit on the Julia set. In this case,
we must consider the effects of critical points contained in
$W^S(\infty)$ and expansion properties of the map $p|J$. We do not
belabor these points here. Whenever we need to choose certain
rays, we will be careful to choose those which behave as we
claim. If the reader is interested in more detail, he should
consult the paper [2] where this topic is treated in the
generality needed here.

Proof. We choose an invariant ray ℓ_1 from infinity to a
fixed point in K_2. To see that such a ray exists suppose that
α_1 and α_2 are the two angles corresponding to the rays from
infinity that limit on the critical point c_1. Then, since the
angles 0 and 1/2 are fixed by $z \mapsto z^3$ and since α_1 and α_2 dif-
fer by 1/3, either 0 or 1/2 must enter B. Moreover, if the
angle of the ray through $p(c_1)$ equals either α_1 or α_2, then it
also equals either 0 or 1/2, and the remaining invariant ray

will enter B and will be disjoint from the forward orbit of c_1.
Consequently, there is always at least one invariant ray from
infinity which limits on a fixed point of K_2.

We define the conjugacy using ℓ_1. Actually, it is best to
first define the inverse map of ϕ. Using f_2^{-1}, we find another
ray ℓ_2 from infinity such that $f_2(\ell_2) = \ell_1$. Then the set
$B - (\ell_1 \cup \ell_2)$ can be written as $U_2' \cup U_3' \cup K_2$ as illustrated in
Figure 7. The map f_2 wraps both U_2' and U_3' entirely around the
slit annulus $U' - (K_2 \cup \ell_1)$. If K is a component of K_p, we
define

$$[\phi^{-1}(K)]_i = \begin{cases} 1 \text{ if } p^i(K) \subset A \\ B \text{ if } p^i(K) = K_2 \\ 2 \text{ if } p^i(K) \subset U_2' \\ 3 \text{ if } p^i(K) \subset U_3'. \end{cases}$$

The definition of ϕ is slightly more involved. Let $V =$
$U' - (K_2 \cup \ell_1)$. Then V is a simply connected domain which does
not contain any critical values of the map p. Therefore we
can define three inverse maps of p, namely

$$I_1: U' \to A, \quad I_2: V \to U_2', \quad \text{and} \quad I_3: V \to U_3' .$$

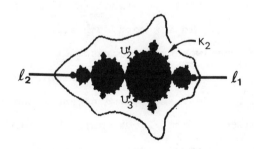

Fig. 7. The decomposition of B into $U_2' \cup K_2 \cup U_3'$ for the
cubic $z^3 - 1.47z + 0.8$.

Since there are essentially two different kinds of sequences
in Σ', we define ϕ in two steps. First, suppose s_n = B and
$s_{n-1} \neq$ B for $s \in \Sigma'$. Then

$$\phi(s) = I_{s_1} \circ I_{s_2} \circ \ldots \circ I_{s_{n-1}} (K_2)$$

Note that s_{n-1} = 1, and consequently, K_2 is a subset of the
domain of definition of $I_{s_{n-1}}$. Secondly, if $s_i \neq$ B for all i,
let s_{n_j} be the subsequence of indices such that s_{n_j} = 1 for
all j. Then let

$$M_j = I_{s_1} \circ \ldots \circ I_{s_{(n_j-1)}} (\bar{A}).$$

and define

$$\phi(s) = \bigcap_{j=1}^{\infty} M_j.$$

In this case, the restrictions of the maps I_1, I_2 and I_3 to $\bar{U}_1$
are all strong contractions in the hyperbolic metric on V.
Therefore, $\phi(s)$ must be a point. □

The second case is that of a periodic kneading sequence
of period 2. In other words, the sequence is BABABA... .
Although we have not worked out all the details in the general
case of a periodic kneading sequence, we expect that the final
result for a periodic sequence will have a similar statement
to the specific case we now consider. Since the kneading
sequence is periodic of period two, we consider three level
sets of h. In addition to L = $h^{-1}(h(c_1))$, we consider p(L)
and $p^{-1}(L)$ (see Figure 8).

Note that $p^{-1}(L)$ consists of two components and bounds
five finite disks. The orbit of c_2 will alternate between two

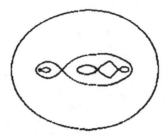

Fig. 8. Three level sets of h and the corresponding
domains that they enclose for the cubic $z^3 - 3.637z + 0.8$.

of them which we label A' and B' where A' ⊂ A and B' ⊂ B. The
map $p^2|B'$ is polynomial-like of degree two, and it has a con-
nected filled-in Julia set which we denote K_B. The map $p|A'$
maps a region K_A homeomorphically on K_B. Since the kneading
sequence is BABABA..., the orbit of c_2 is contained in the
union $K_A \cup K_B$. Our symbolic coding in this case is based on
the following geometric decomposition of U'. Take a ray ℓ_1
from infinity whose angle is periodic of period two under
$z \to z^3$ on S^1 and which limits onto a periodic point of period
two in K_A. Then the inverse $(p|B)^{-1}$ applied to ℓ_1 yields two
other rays from infinity ℓ_2 and ℓ_3. With these rays, we get a
decomposition of U' (see Figure 9) which is similar to the
decomposition we used in Theorem 4 and which gives our symbolic coding. If $V = U' - (K_A \cup \ell_1)$, then we have three inverse
maps I_1, I_2, and I_3 defined on U', V, and V respectively. The
construction of the coding proceeds in a similar manner to the
proof of Theorem 4 with a slightly different symbol space Σ'
which is now a subspace of $\Pi\{1,2,3,A,B\}$. However, before we
give a precise description, we introduce a bit of terminology
which simplifies the definition. Given a sequence $s \in \Pi\{1,2,3\}$,
its associated A-B sequence is gotten by replacing 1 by A and

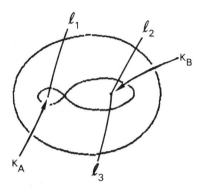

Fig. 9. The sets K_A and K_B and the rays from infinity
which define the partition of U' for the cubic $z^3 - 3.637z + 0.8$.

2 and 3 by B. A sequence s ϵ Σ' for this kneading sequence if
and only if

(1) $s_k = B \Rightarrow s_{k+1} = A$,

(2) $s_k = A \Rightarrow s_{k+1} = B$,

(e) B and 1 are the only symbols that can precede A,

(4) A, 1, and 2 are the only symbols that can precede B, and

(5) if s ϵ Σ_3, then its associated A-B sequence cannot end

with the sequence $BA\overline{BA}$.

Theorem 5. Suppose the A-B kneading sequence is BABABA...
and Σ' is the σ-invariant subspace of Σ_5 defined above. Then
there exists a conjugacy σ: Σ' $\rightarrow$ {components of K_p}. Those
components which are images of sequences not containing the
symbols A or B are points.

The final case which we consider is the kneading sequence
BAAA... . We call this sequence preperiodic because, although
the sequence is not periodic, it is eventually periodic (under
the shift map applied to kneading sequences). This case is
quite different from the other two considered because the

filled-in Julia set K is totally disconnected. However, K

(which equals J) is not topologically conjugate to $\sigma | \Sigma_3$

because it contains a fixed point with only two distinct pre-

images rather than three. Although we construct a symbolic

coding for this case, its topological properties are quite dif-

ferent from those used in the proof of Theorem 4. We should

also note that this case was first studied by Brolin [6,

Theorem 13.8] to provide a counterexample to a conjecture of

Fatou regarding critical points which are contained in the

Julia set. Brolin proved that J was totally disconnected.

R. Devaney actually pointed out how the techniques described

in this section can be used to study this case. As always, we

start with the level set $L = h^{-1}(h(c_1))$ which is a pinched

circle. The fact that this kneading sequence is BAAA...

implies that $p(c_2)$ is the repelling fixed point α which is the

only member of $K_{p|A}$. First, note that the component of K con-

taining α is the singleton $\{\alpha\}$ because that component is con-

tained in the nested intersection $\cap_{k=0}^{\infty}(f_1)^{-1}(A)$. The boun-

daries of those sets are not in K, and they contract to a

point. Now we choose a ray from infinity ℓ which limits on α.

Let V be U' - ℓ (see Figure 10). On V, there exist three

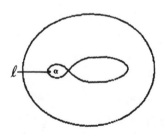

Fig. 10. The decomposition of U' induced by the ray ℓ for
the cubic $z^3 - 4.03474z + 0.8$.

inverse functions I_1, I_2, and I_3 of the map p which map V into
disjoint subsets of V. We can now characterize every component
(and every point in K) just as we have done above. Let Σ' be
the subset of Σ_3 consisting of all sequences which contain
either infinitely many 2's or 3's. Then there exists a conjugacy ϕ from Σ' to the set of components of K which do not
eventually map onto α. It is defined by

$$\phi(\{s_i\}) = \bigcap_{k=1}^{\infty} I_{s_0} \circ I_{s_1} \circ \ldots \circ I_{s_k} (V),$$

and one can prove that each such component is actually a point
because these holomorphic functions are strict contractions on
B in the Poincaré metric on V. The only remaining components
of K which are left to analyze are the ones which are eventually mapped to α. Therefore, they are $\{\alpha\}$, $\{c_2\}$, and all
the singletons corresponding to the inverse images of c_2.
Therefore, we have the following theorem.

Theorem 6. If the A-B kneading sequence for the cubic
p(z) is BAAA..., then the map $p|J$ is topologically conjugate
to the quotient $(\Sigma_3/\sim)$ where $\sim$ is the smallest equivalence
relation which is generated by the identification 2111... $\sim$
3111... and which yields a space on which the quotient
$\tilde{\sigma}: (\Sigma_3/\sim) \to (\Sigma_3/\sim)$ is well-defined.

3. SUMMARY AND OPEN PROBLEMS

    In this paper, we have introduced a kneading sequence with
two symbols as a device to help in our study of the dynamics
of cubics with one critical point iterating to infinity and
with another critical point whose orbit is bounded. We were

then able to explicitly construct conjugacies of the component space of the filled-in Julia sets to manageable symbol
spaces for three different types of kneading sequences --
BBB...., BABABA..., and BAAA... . We expect that these con-
structions indicate how symbolic codings should be constructed
when the kneading sequence is periodic or preperiodic. We
conclude with three unresolved questions which are fundamental
to this study.

   Problems. (1) Can the A-B kneading sequence be aperiodic?
If so, what is the associated dynamics?
(2) If the kneading sequence is preperiodic, is the Julia set
a Cantor set?
(3) For a given kneading sequence, calculate the number of its
Mandelbrot sets in each leaf of the trefoil clover.

REFERENCES

[1]   Blanchard, P., Complex Analytic Dynamics on the Riemann
      Sphere, Bull. Amer. Math. Soc. (New Series) 11 (1984),
      85-141.

[2]   Blanchard, P., Symbols for Cubics and Other Polynomials,
      preprint.

[3]   Branner, B. and Hubbard, J., Iteration of Complex Cubic
      Polynomials I: The Global Structure of Parameter Space,
      personal communication.

[4]   Branner, B. and Hubbard, J., Iteration of Complex Cubic
      Polynomials II: Patterns and Parapatterns, personal
      communication.

[5]   Branner, B., The Parameter Space for Complex Cubic Poly-
      nomials, Proceedings of the Conference on Chaotic
      Dynamics, Georgia Tech, 1985.

[6]   Brolin, H., Invariant Sets Under Iteration of Rational
      Functions, Arkiv für Matematik 6 (1965), 103-144.

[7] Douady, A., Systèmes Dynamiques Holomorphes, Séminaire
     Bourbaki, 1982/1983, Exposé 599, Astérisque 105-106
     (1983), Societé math. de France.

[8] Douady, A. and Hubbard, J., On the Dynamics of
     Polynomial-Like Mappings, Ann. Scient. Ec. Norm. Sup.,
     to appear.

[9] Douady, A. and Hubbard, J., Itération des polynômes
     quadratiques complexes, C. R. Acad. Sci. Paris 294
     (1982), 123-126.

[10] Douady, A. and Hubbard, J., Étude Dynamique des
     Polynômes Complexes, Publications Mathématiques D'Orsay,
     Université de Paris-Sud.

[11] Fatou, P., Sur les équations fonctionelles, Bull. Soc.
     Math. France 47 (1919), 161-271; 48 (1920), 33-94, and
     208-314.

[12] Guckenheimer, J., Endomorphisms of the Riemann Sphere,
     Proc. Sympos. Pure Math. 14 (S. S. Chern and S. Smale,
     eds.), Amer. Math. Soc. 1970, 95-123.

[13] Julia, G., Memoire sur l'itération des fonctions
     rationnelles, J. Math. pures et app. 8 (1918), 47-245.
     See also Oeuvres de Gaston Julia, Gauthier-Villars,
     Paris 1, 121-319.

[14] Jakobson, M., Structure of Polynomial Mappings on a
     Singular Set, Mat. Sb. 80 (1968), 105-124; English
     transl. in Math. USSR-Sb. 6 (1968), 97-114.

[15] Jakobson, M., On the Problem of the Classification of
     Polynomial Endomorphisms of the Plane, Mat. Sb. 80
     (1969), 365-387; English transl. in Math. USSR-Sb. 9
     (1969), 345-364.

[16] Mandelbrot, B., The Fractal Geometry of Nature, Freeman
     1982.

[17] Mandelbrot, B., Fractal Aspects of the Iteration of
     $z \mapsto \lambda z(1-z)$ for complex λ and z, Ann. New York Academcy
     of Sci. 357 (1980), 249-259.

[18] Thurston, W., On the Dynamics of Iterated Rational Maps,
     preprint.

CALCULATION OF TAYLOR SERIES FOR JULIA

SETS IN POWERS OF A PARAMETER*

W. D. Withers

Department of Mathematics
U. S. Naval Academy
Annapolis, MD

ABSTRACT

Let $\{f_c\}$ be a family of mappings dependent on a parameter c. A method is presented for calculating sets which can represent the successive derivatives with respect to c of the Julia set for f_c at some value c_0 of c. In practice, the Taylor series constructed from these derivatives yields good approximations to the Julia sets for f_c for c in a neighborhood of c_0.

1. INTRODUCTION

In this article we present methods for representing the Julia sets of a family of maps dependent on a parameter c by a Taylor series in powers of $(c-c_0)$. As seen in Section 3, the associated Taylor polynomials provide in practice quite good approximations to the Julia set for parameter values c reasonably near c_0. Use of the Taylor polynomials allows fast calculation of pictures of Julia sets for several different values

---

*This work was supported by a grant from the Naval Academy Research Council.

of c in a neighborhood of c_0.

We will treat as an example the family of quadratic maps $f_c(z) = z^2 - c$, but our methods can easily be extended to a wide variety of families of maps. For the family of quadratic maps the structure of the Julia set is well understood in terms of the position of c relative to the Mandelbrot set M in parameter space; for example, see A. Douady and J. H. Hubbard [2].

We now review a few facts concerning Julia sets for the quadratic map $f(z) = z^2 - c$. Let S be the set of sequences $\{(a_1, a_2, \ldots)\}$ where $a_n = \pm 1$. Let $\sqrt{c + z}$ denote one branch of the inverse of $f_c(z) = z^2 - c$; then $-\sqrt{c + z}$ is the other branch. Choose $s = (a_1, a_2, \ldots) \in S$. Then for almost every $z \in C$,

$$\lim_{n \to \infty} a_1 \sqrt{(c + a_2 \sqrt{(c + \ldots + a_n \sqrt{(c + z)})})} \text{ exists,}$$

and takes a value independent of z; we will denote this number by $\beta_c(s)$. The range of the function β_c over S is the Julia set J_c for $f_c(z) = z^2 - c$. Moreover, if $\sigma: S \to S$ is the shift opertor, $\sigma(a_1, a_2, \ldots) = (a_2, a_3, \ldots)$, then the action of σ on S is equivalent to the action of f_c on J_c; i.e.,

$$f_c(\beta_c(s)) = \beta_c(\sigma(s)).$$

Depending on c, the function β_c may not be injective, but that need not concern us here.

2. DERIVATIVES OF JULIA SETS WITH RESPECT
    TO THE PARAMETER

We now consider the dependence of $\beta_c(s)$ on c for fixed s; in particular, we ask whether $\beta_c(s)$ is differentiable with respect to c. The branch cuts for $\pm\sqrt{c + z}$ may be chosen dependent on c so that $\beta_c(s)$ varies continuously with c. If for each $s \in S$ the function $\beta_c(s)$ is differentiable with respect to c in a neighborhood of c_0 independent of S, then we may consider the set J_c to be differentiable with respect to c in that neighborhood. If for each s we find a Taylor series for $\beta_c(s)$ in powers of $(c - c_0)$, then we have found in effect a Taylor series for J_c in powers of $(c - c_0)$.

M. F. Barnsley, J. S. Geronimo and A. N. Harrington [1] have proven rigorously that $\beta_c(s)$ is an analytic function of c for $|c| < 1/4$, and hence that it has a Taylor series in powers of $(c - c_0)$ for $|c_0| + |c - c_0| < 1/4$. To find the Taylor series for $\beta_c(s)$ we must evaluate the successive derivatives $D\beta_c(s)$, $D^2\beta_c(s), \ldots$, at $c = c_0$. (We let D denote differentiation with respect to C.)

We now introduce subscript notation for iterates. For fixed $s \in S$, we write $z_0(c)$ or z_0 for $\beta_c(s)$; then we denote $f_c(z_0) = \beta_c(\sigma(s))$ by $z_1(c)$ or z_1, $f_c(z_1) = \beta_c(\sigma^2(s))$ by $z_2(c)$ or z_2, and so on.

Recall that β_c satisfies the functional equation

$$\beta_c(\sigma(s)) = f_c(\beta_c(s)) = \beta_c(s)^2 - c.$$

Differentiating both sides with respect to c, we obtain

$$D\beta_c(\sigma(s)) = 2\beta_c(s)D\beta_c(s) - 1.$$

We thus have a functional equation in $D\beta_c$ and β_c. In our iterative notation, this becomes

$$Dz_1 = 2z_0Dz_0 - 1. \qquad (1)$$

Solving for Dz_0, we have

$$Dz_0 = \frac{1}{2z_0} + \frac{1}{2z_0} \, Dz_1. \tag{2}$$

But it is also true that $Dz_1 = \frac{1}{2z_1} + \frac{1}{2z_1} \, Dz_2$, etc.; thus

$$Dz_0 = \frac{1}{2z_0} + \frac{1}{2z_0} \left(\frac{1}{2z_1} + \frac{1}{2z_1} \, Dz_2 \right)$$

$$= \frac{1}{2z_0} + \frac{1}{2z_0} \left(\frac{1}{2z_1} + \frac{1}{2z_1} \left(\frac{1}{2z_2} + \frac{1}{2z_2} \, Dz_3 \right) \right)$$

etc.

$$= \sum_{k=0}^{\infty} \prod_{n=0}^{k} \frac{1}{2z_n} \tag{3}$$

If the series (3) converges, then it gives Dz_0 in terms of z_0 and its iterates. Note also that it allows us to calculate Dz_0 without s such that $z_0 = \beta_c(s)$.

When does the series (3) converge? Let us introduce the concept of a _Lyapunov exponent_. Suppose for $z_0 \in J_c$,

$$\lim_{N \to \infty} \frac{1}{N} \sum_{n=1}^{N} \log |f_c'(z_n)| \quad \text{exists};$$

then we call this limit the Lyapunov exponent of z_0 and denote it by $\lambda(z_0)$. We use $f_c'(z)$ to denote $\frac{\partial}{\partial z} f_c(z)$ as opposed to $Df_c(z) = \frac{\partial}{\partial c} f_c(z)$. Note that $\lambda(z_0) = \lambda(z_n)$ for all $n \geq 0$. It is not difficult to see that if $\lambda(z_0) > 0$, then the series (3) converges. If c is on the boundary of M, there exists $z_0 \in J_c$ with $\lambda(z_0) = 0$; cf [2]. On the other hand, it can be shown that if $|c| > 17/4$, then $\lambda(z_0) > 0$ for any $z_0 \in J_c$. This suggests that $\beta_c(s)$ is analytic in c for all $|c| > 17/4$ (in addition to $|c| < 1/4$ as shown in [1]), and perhaps for all c except on the boundary of M.

We now return to (1) and differentiate with respect to c:

$$D^2\beta_c(\sigma(s)) = 2\beta_c(s) + (D\beta_c(s))^2.$$

We now have a functional equation in β_c, $D\beta_c$, and $D^2\beta_c$. In iterative notation it becomes

$$D^2 z_1 = 2z_0 D^2 z_0 + (Dz_0)^2.$$

As before, we solve for $D^2 z_0$:

$$D^2 z_0 = \frac{-1}{2z_0}(Dz_0)^2 + \frac{1}{2z_0}D^2 z_1; \qquad (4)$$

and derive a series expansion for $D^2 z_0$:

$$D^2 z_0 = \sum_{k=0}^{\infty} -(Dz_k)^2 \prod_{n=0}^{k} \frac{1}{2z_n}.$$

This series will converge whenever $\lambda(z_0) > 0$ and Dz_0 is bounded independently of z_0 (i.e., $\beta_c(s)$ is bounded independently of s).

In this fashion, for any $m > 1$ we can obtain a functional equation in $\beta_c, D\beta_c, \ldots, D^m\beta_c$:

$$D^m\beta_c(\sigma(s)) = \sum_{k=0}^{m} \binom{m}{k} D^k\beta_c(s) D^{m-k}\beta_c(s),$$

and a corresponding series expression for $D^m z_0$ in terms of $z_n, Dz_n, \ldots, D^{m-1}z_n$; $n = 0,1,2,\ldots$:

$$D^m z_0 = \sum_{k=0}^{\infty} \left(- \sum_{n=1}^{m-1} \binom{m}{n} D^{m-n} z_k D^n z_k\right) \prod_{n=0}^{k} \frac{1}{2z_n}$$

which converges provided $\lambda(z_0) > 0$ and $Dz_0, D^2 z_0, \ldots, D^{m-1}z_0$ are bounded independently of z_0.

3. RESULTS OF CALCULATIONS

One method for making pictures of Julia sets is to choose z_0 almost anywhere in C and then calculate and plot $z_{-1}, z_{-2}, \ldots$

from the formula

$$z_{n-1} = \pm\sqrt{c + z_n} \; , \tag{5}$$

choosing a branch at random each time. This process is similar to our definition of $\beta_c(s)$ where s is a random sequence. If $z_0 \in J_c$, then the set of backward iterates $z_{-1}, z_{-2}, \ldots$ will be dense in J_c with probability 1. We may think of this process as beginning by estimating z_0 with a large error. The repelling nature of J_c insures that the error in the computed value of z_n decreases to the size of computation error as $n \to -\infty$.

This method is easily extended to the calculation of derivative sets for J_c with respect to c. We estimate the values of $z_0, Dz_0, \ldots, D^m z_0$, with perhaps large errors and then iterate backwards using equations (5), (2), (4), etc.:

$$z_{n-1} = \pm\sqrt{c + z_n} \; ;$$

$$Dz_{n-1} = \frac{1}{2z_{n-1}} + \frac{1}{2z_{n-1}} Dz_n ;$$

$$D^2 z_{n-1} = \frac{-1}{2z_{n-1}} (Dz_{n-1})^2 + \frac{1}{2z_{n-1}} D^2 z_n ;$$

etc.

Provided the average value of $\lambda(z_0)$ (with respect to a measure which we do not detail here) is positive, the errors in the calculated values decrease to the size of computation errors as $n \to -\infty$.

Figures 1-3 show results of calculations involving values of c inside the cardioid $r = 1/2 \, (1 + \cos \theta)$, where $c = re^{i\theta} - 1/4$. The interior of this cardioid is the connected component of the interior of M which contains $c = 0$. In Figure 1(a) and (b) we

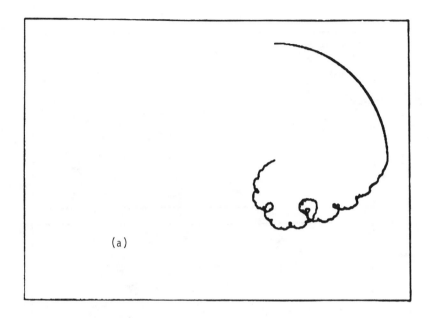

(a)

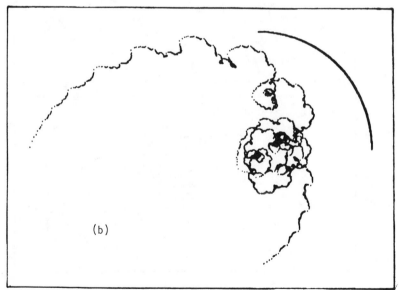

(b)

Fig. 1.   (a) First- and (b) second-derivative curves for
the Julia set at c = 0, with the Julia set itself shown for
comparison.   For clarity, only values corresponding to z_0 in
the first quadrant are shown.

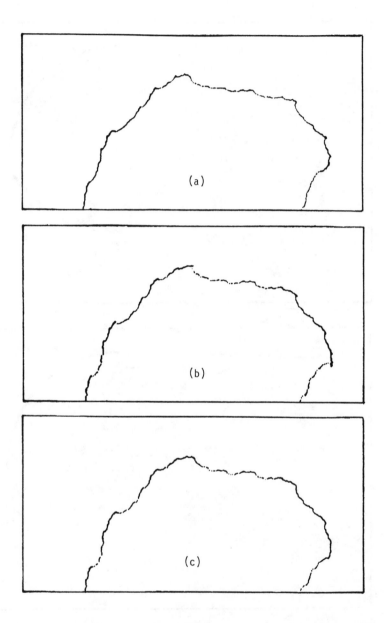

Fig. 2.   (a) First- and (b) second-order approximations to the Julia set at c = .3i, taken from the point $c_0 = 0$.   (c) The actual Julia set at c = .3i.   The curves in (a) and (b) are linear combinations of the curves in Figure 1.

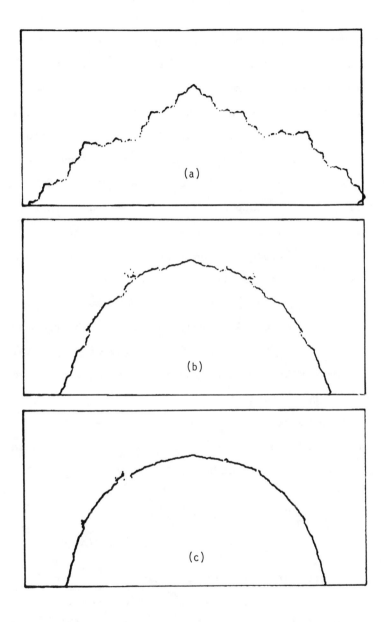

Fig. 3. (a) The Julia set for $c = .5 + .1i$. (b) First-order and (c) second-order approximations to the Julia set at $c = 0$ from $c_0 = .5 + .1i$.

we have the first and second-order derivative sets, respectively, for J_c at $c = 0$. At $c = 0$, J_c is the unit circle; it
is also shown for comparison. For clarity, only values corresponding to values of z_0 in the first quadrant are shown;
all the curves are symmetric about the real and imaginary
axes. Thus Fig. 1(a) shows the values taken by $D\beta_c(s)$ and
Fig. 1(b) shows those taken by $D^2\beta_c(s)$ as $\beta_c(s)$ takes values
in the first quadrant of the unit circle. When $z_0 = 1$, $Dz_0 = 1$
and $D^2z_0 = -2$; when $z_0 = i$, $Dz_0 = 0$ and $D^2z_0 = -i$. The sets
appear to be intricate curves; it can be shown that the derivative set at any order at $c = 0$ is a curve of Hausdorff dimension 1. It follows that the Taylor-polynomial approximation
of any order to the Julia set J_c from $c_0 = 0$ is a curve of
Hausdorff dimension 1. This means that Hausdorff dimension is
not among the properties of J_c approximated by the Taylor polynomial from $c_0 = 0$ and also that the Taylor polynomial from
$c_0 = 0$ cannot be a good approximation to J_c when $c \in M$ and J_c
is therefore disconnected.

In Figure 2(a) and (b) we see the first- and second-order
approximations to J_c at $c = .3i$ from $c_0 = 0$ as calculated from
the first- and second-order Taylor polynomial formulas.

$$z_0(c) \simeq z_0(c_0) + (c - c_0)Dz_0(c_0);$$

$$z_0(c) \simeq z_0(c_0) + (c - c_0)Dz_0(c_0) + 1/2(c - c_0)^2D^2z_0(c_0).$$

Thus the curves in Fig. 2(a) and (b) are linear combinations of the curves in Fig. 1. Once the curves in Fig. 1 have
been calculated, a similar approximation can be made for any c
near $c_0 = 0$ by taking the appropriate linear combination.
Fig. 2(c) shows the actual Julia set calculated for $c = .3i$

shown for comparison. Figures 2 and 3 show the region of the
complex plane $-1.3 \le \text{Re } z \le 1.3$, $0 \le \text{Im } z \le 1.3$; all the sets
involved are symmetric about the origin.

Figure 3(a) shows the Julia set calculated at $C = .5 + .1i$.
Figure 3(b) and (c) show the first- and second-order approximations to the Julia set at $c = 0$ from $c_0 = .5 + .1i$. Considering
the position of $.5 + .1i$ within the cardioid, the value $c = 0$
should be well outside the radius of convergence of the Taylor
series; nonetheless we obtain fairly good approximations to the
unit circle.

REFERENCES

[1] M. F. Barnsley, J. S. Geronimo, and A. N. Harrington, On
    the invariant sets of a family of quadratic maps, Commun.
    Math. Phys. 88 (1983), 479-501.

[2] A. Douady and J. H. Hubbard, Iteration des polynomies
    quadratiques complexes, C. R. Acad. Sc. Paris, Serie I.
    294 (1982), 123-126.

DIOPHANTINE PROPERTIES OF JULIA SETS

Pierre Moussa

Service de Physique Théorique
Centre d'Etudes Nucléaires de Saclay
F-91191 Gif-sur Yvette Cedex

ABSTRACT

We show that algebraic integers included with all conjugates in the filled-in Julia set of a polynomial with integer coefficients are preperiodic points for the polynomial
transformation. Therefore they must have a finite forward
orbit. Various extensions of this result will be discussed:
polynomial with coefficients in an algebraic number ring,
quantitative formulation and connection with Lehmer's problem,
analogous properties for the Mandelbrot set in the parameter
complex plane.

1. INTRODUCTION

This talk covers results obtained in collaboration with
J. S. Geronimo and D. Bessis [1]. I also benefited from dis-
cussions with A. Douady, F. Gramain and J. P. Kahane. I thank
J. M. Luck for his help in preparing the pictures.

A classical problem in number theory is the following question:

Let D be a compact subset of the complex plane. What are

the algebraic integers which belong to D as well as all their

conjugates? In other words, what are the monic polynomials

with rational integer coefficients which have all their roots

in D?

We first met this problem in a previous work [2,3] on the

diophantine moment problem: find measures, with support on the

real line or in the complex plane, such that all moments are

integers. Let D be the support. To find discrete measures

with integer moments require to look for algebraic integers

sitting in D with all their conjugates, and point masses of

the discrete measure must been placed at every conjugates.

The classical problem of arithmetic entire function [4,5,

6] gives another example: we look for entire holomorphic

functions such that E(z) is a rational integer when z is a

natural integer. Superposition of exponentials may belong to

this class: $E(z) = \sum_i a_i \exp(b_i z)$ provided that each $\exp(b_i)$

is an algebraic integer, and that the sum includes also the

conjugates. Growth condition on E(z) for $|z|$ large will deter-

mine the domain to which all $\exp(b_i)$ must belong.

2.   FEKETE'S THEOREM, KRONECKER'S THEOREM

Fekete showed in 1923 [7] that if the domain D has a

transfinite diameter less than one, there is only a finite

number of algebraic integers sitting in D together with their

conjugates. This result allows to get a better understanding

of the following classical results by Kronecker:

a)   Let D be the closed unit disk. Let z be an algebraic

integer such that z and all its conjugates belong to D. Then

either z = 0, or z is a root of unity.

b)   Let D be the closed interval [-2,+2].   Let z be an
algebraic integer such that z and all its conjugates belong to
D.   Then z is of the form: z = 2 cos πq, with q a rational
number.

In both examples, the transfinite diameter of D is just
equal to one. The solution of the problem is not known for a
general domain, for example when D is the image of a disk
through the exponential map [4,5].   However, we shall see in
the next section that Kronecker's results are just particular
cases of domains invariant under a polynomial transformation.

3.   JULIA SETS ARE NATURAL SETS FOR LOCALIZATION OF ALGEBRAIC INTEGERS

Let T be a polynomial of degree at least 2 with integer
coefficients. We define its filled-in Julia set K_T as the
complement of the basin of attraction of the point at infinity
under the polynomial map T [8].   We then have our main result
[1]:

Theorem 1.   Let T be a polynomial with rational integer
coefficients. Any algebraic integer included in K_T together
with its conjugates is a preperiodic point for T. If in addition T is monic, then all preperiodic points for T are algebraic integers.

We recall first the definition of preperiodic points: the
complex number z is said to be preperiodic for T, if there
exist integers k > 0 and $\ell \geq 0$ such that:

$$T^{(k+\ell)}(z) = T^{(\ell)}(z)$$

where $T^{(0)}(z) = z$ and $T^{(n)}(z)$ is recursively defined by
$T^{(n)}(z) = T(T^{(n-1)}(z))$. All preperiodic points of T belong to
K_T, and it is easy to see that when T is monic, all preperiodic
points are algebraic integers. The conjugates are also preperiodic. Let us now check the first statement of theorem.
Assume T is a polynomial with integer coefficients (non-
necessarily monic). Let z be an algebraic integer (of degree
r) sitting in K_T together with its conjugates. By definition
of K_T, all iterates $T^{(n)}(z)$, $n \geq 0$, belong to K_T, and they are
algebraic integers of degree at most r. Their conjugates are
also iterates of conjugates of z, and they also belong to K_T.
However there is only a finite number of algebraic integers of
bounded degree, sitting with their conjugates in a bounded
set, since then all symmetric functions are also bounded and
can take only integer values. However K_T is a bounded set,
therefore the sequence $T^{(n)}(z)$ can only take a finite number of
distinct values, and two of them (at least !) must be equal.
This achieves the proof. Notice that once more we have a
limiting example of Fekete's theorem, since the transfinite
diameter of K_T is equal to one when T is a monic polynomial
[8,10].

We recover the above mentioned Kronecker's results:

Case (a): For $T = z^2$, since any root of unity is preperiodic for $T = z^2$.

Case (b): For $T = z^2 - 2$, since $z = 2 \cos \pi q$, is preperiodic for $T = z^2 - 2$ when q is rational.

The set K_T can be connected or infinitely disconnected [8].
When the degree of T is equal to 2, the set K_T is connected
in the following cases displayed in Figure 1. T is equal to

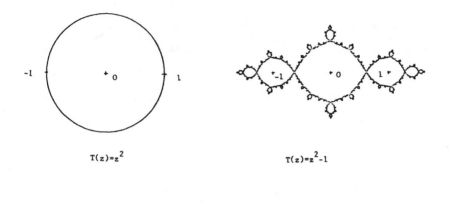

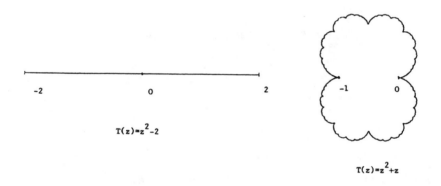

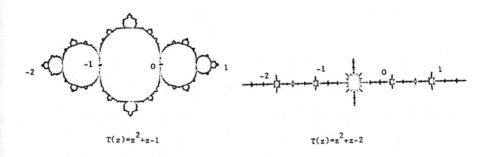

Fig. 1. Some diophantine Julia sets: the six connected cases in degree two.

z^2, $z^2 - 1$, $z^2 - 2$, $z^2 + z$, $z^2 + z - 1$, or $z^2 + z - 2$. K_T is
also connected when T is deduced from one of the previous six
cases denoted $\tilde{T}$, through the transformation: $T(z) = \tilde{T}(z - n) + n$,
where n is any rational integer. The set K_T and $K_{\tilde{T}}$ are then
just deduced from one another through a translation $z \to z + n$.
For any other quadratic T, K_T is a Cantor set [9,10].

We have now obtained a number theoretical characterization
of preperiodic points, that is points with a finite orbit under
a polynomial transformation with integer coefficients. In the
next section we shall obtain analogous results for a larger
class of polynomials.

4. GENERALIZATION TO POLYNOMIALS WITH ALGEBRAIC INTEGERS COEFFICIENTS

Let θ be an algebraic integer of degree s, and $\theta_1 = \theta$,
$\theta_2, \ldots, \theta_s$ its conjugates. We denote $A[\theta]$ the set of linear
combinations $\sum_{j=0}^{s-1} p_j \theta^j$ with rational integer coefficients p_j.
$A[\theta]$ is a ring and we define in the same way the conjugate
rings $A[\theta_i]$ for $i = 1, \ldots, s$. We first note that if $x \in A[\theta]$,
x is itself an algebraic integer of degree at most s, the conjugates of which belong to the corresponding $A[\theta_i]$.

Now we shall say that z is algebraic integer over $A[\theta]$ if
z is a solution of an equation $P(z) = 0$ where P is a monic
polynomial with coefficients in $A[\theta]$. If P is not expressible
as a product of monic polynomials of lower degree, with coefficients in $A[\theta]$, the degree of P is the degree of the algebraic integer z over $A[\theta]$. The other roots of P are the conjugates of z. We shall call then internal conjugates. By
conjugating the coefficients of P, we get the conjugate

polynomials P_i, $i = 1,2,\ldots,s$, $P = P_1$, and since $\prod\limits_{i-1}^{s} P_i$ is a

polynomial with integer coefficients, z is also an ordinary

algebraic integer of degree (rs), and it has therefore $r(s-1)$

other conjugates, solutions of $P_i = 0$ for $i = 2,\ldots,s$. We

shall call the latter external conjugates.

Now consider a polynomial T with coefficient in $A[\theta]$. By

conjugating its coefficients, we also define the conjugates

polynomials $T_1 = T, T_2, T_3, \ldots, T_s$. Theorem 2 will allow us to

characterize the preperiodic points for T.

Theorem 2. Let z be an algebraic integer over $A[\theta]$ such

that z and its internal conjugates belong to K_T, and such that

its external conjugates belong to the corresponding K_{T_i}, then

z is preperiodic for K_T. If in addition T is monic, all pre-

periodic points of T are algebraic integers over $A[\theta]$, with

conjugates belonging to the corresponding K_{T_i}.

Once one has observed that under the assumptions, the

external conjugates of the iterates $T^{(n)}(z)$ remain also bounded,

the proof follows the same argument as for Theorem 1. Our

argument shows that it is convenient to relate Julia sets

associated to conjugate polynomials, and it would be interesting

to analyze the corresponding shapes of conjugates Julia sets.

5. QUANTITATIVE FORMULATION AND GENERALIZATION
   OF LEHMER'S PROBLEM

A. Circle Case: Lehmer's Question

Let us consider the algebraic integer z, with conjugates

$z = z_1, z_2, \ldots, z_r$. We call $M(z)$ the Mahler's measure of z,

defined by

$$M(z) = \prod_{i=1}^{r} \sup(1, |z_i|)$$

Kronecker's theorem is equivalently rephrased as: $M(z) = 1$ if and only if $z = 0$ or z is a root of unity. We can refine this statement by bounding the degree r. We then have the statement: $\exists \, \varepsilon(r) > 0$ such that we have $M(z) < 1 + \varepsilon(r)$ for an algebraic integer z with degree at most r, if and only if $z = 0$ or z is a root of unity. There is a long story about improving the bound $\varepsilon(r)$, the best available at the present time is [11]: $\varepsilon(r) = C(\ln(\ln r)/\ln r)^3$. Lehmer's question is: does it exist an ε independent of r such that $M(z) > 1 + \varepsilon$ for any non-vanishing algebraic integer which is not a root of unity [12].

B. A Generalized Measure for Algebraic Integers

Given the filled-in Julia set K_T, we need a way of measuring how far from K_T is a complex number z. For this we use the following procedure:

Let $T(z)$ be a monic polynomial of degree $d \geq 2$:
$$T = z^d + \sum_{i=1}^{d} t_i z^{d-i} \, .$$

Let τ be defined as $\tau = \sup_i |t_i|$. Then we have the following lemmas which are easily proven:

Lemma 1. Given ε, ε', $1 > \varepsilon > 0$, $\varepsilon' > 0$ setting $\delta = 1/(1-d)$ and choosing R as

$$R = \sup\left(\frac{\tau d}{\varepsilon}, \frac{\tau d}{\varepsilon'}, (1-\varepsilon)^\delta\right), \text{ then:}$$

$$|z| > R \Rightarrow |z|^d(1+\varepsilon') > |T(z)| > |z|^d(1-\varepsilon) > R.$$

Lemma 2. Given α, α', $1 > \alpha > 0$, $\alpha' > 0$ and choosing ε and ε' as:

$$\varepsilon' = (1+\alpha')^{d-1} - 1, \ \varepsilon = 1 - (1-\alpha)^{d-1}, \text{ and R as in}$$

Lemma 1, then:

$$k > 0, \ |z| > R \Rightarrow (1+\alpha')^{d^k} > |T^{(k)}(z)/z^{d^k}| > (1-\alpha)^{d^k}.$$

Lemma 3. We define $\beta_T(z)$ as follows: if $z \in K_T$, $\beta_T(z) = 1$, and if $z \notin K_T$, $\beta_T(z) = \lim_{k\to\infty} |T^{(k)}(z)|^{1/d^k}$ (this limit exists). Then we have:

(i) $\beta_T(T(z)) = (\beta_T(z))^d$;

(ii) $\beta(z) > 1$ if and only if $z \notin K_T$;

(iii) $|z| > R \Rightarrow (1+\alpha') > \beta(z)/|z| > (1-\alpha)$, where α, α', R as in Lemma 2;

(iv) choosing then $k_T = \sup(R, (1-\alpha)^{-1})$, we have for any z, $|z| < k_T \beta_T(z)$.

Given an algebraic integer z, with conjugates $z_1 = z, z_2, \ldots, z_r$, we define its generalized measure $M_T(z)$ as:

$$M_T(z) = \prod_{i=1}^{r} \beta_T(z_i).$$

Then our Theorem 2 says that if T is a monic polynomial with integer coefficients, $M_T(z) = 1$ if and only if the algebraic integer z is preperiodic for T.

We recover the usual Mahler's measure for $T = z^2$, since in this case $\beta_T(z) = \sup(|z|, 1)$.

C. The Generalized Lehmer's Problem

Now we consider an arbitrary algebraic integer of degree at most r, with a bounded measure: $M_T(z) < m$. So we have $\beta_T(z) < m$ and similarly $\beta_T(z_i) < m$ for any conjugate z_i of z.

We then consider the finite sequence $z, T(z), \ldots, T^{(k)}(z)$,
which contains $(k + 1)$ different algebraic integers if z is not
preperiodic for T. According to Lemma 3(i) any number y in
this sequence satisfies $\beta_T(y) < m^{d^k}$, and the same inequality
holds if we replace y by one of its conjugates. Therefore
according to Lemma 3(iv), there is at least $(k + 1)$ distincts
algebraic numbers of degree at most r in the disk of radius
$k_T m^{d^k}$, together with their conjugates. However, it is easy to
bound the number of algebraic integers of degree at most r
sitting with their conjugates in a disk of radius $\rho > 1$, since
the pth symmetric function of the roots is bounded in modulus
by $\rho^p\binom{r}{p} < \rho^p 2^r$ and can take at most $1 + 2\rho^p 2^r < 3\rho^p 2^r$ distinct
values. Hence the number of algebraic integers is at most:

$$\prod_{i=1}^{r} 3\rho^i 2^r = 3^r 2^{r^2} \rho^{\frac{r(r+1)}{2}} .$$

Then if z is not preperiodic for T, for any k we must have
$k + 1 < Ae^{Ld^k}$, with

$$A = 3^r 2^{r^2} k_T^{\frac{r(r+1)}{2}} = A(r), \quad \text{and} \quad L = \frac{r(r+1)}{2} \ln m.$$

A rather tedious argument shows that if $0 < L < \dfrac{1}{A \, d^{e^A} \ln d}$,
the inequation $k + 1 > Ae^{Ld^k}$ will have positive integer solution and z has to be preperiodic. This shows the following
result:

Theorem 3. If z is an algebraic integer of degree at most
r, which is not preperiodic for the monic polynomial T with
integer coefficients, then we have $M_T(z) > m(r)$, with

$$\ln m(r) = \frac{2d^{-eA(r)}}{r(r+1)A(r) \ln d} ,$$

with $\ln A(r) = r \ln 3 + r^2 \ln 2 + \frac{r(r+1)}{2} \ln k_T$.

Of course this bound being very general and explicit, remains a very loose one, and there is much work to do to get an estimate comparable to the circle case.

Nevertheless, the Lehmer's question still holds: Is the best possible $m(r)$ bounded from below by a constant strictly greater than 1, and independent on the degree r?

6. EXTENSION TO THE MANDELBROT M SET

Mandelbrot's M set for the family $f_c(z) = z^2 + c$ is defined as the set of the complex values of c such that the filled-in Julia set K_c associated to f_c is connected [9]. It is equivalently defined by the set of values of c such that the point $z = 0$ belongs to K_c. From the arguments given by Douady and Hubbard [10], one can deduce that the transfinite diameter of M is equal to 1. One easily sees that when $c \in M$, the set K_c is uniformly bounded (certainly in a disk of radius smaller than 4). As pointed out by Douady [13], this allows one to determine all algebraic integers sitting in M together with their conjugates. We have the following result:

Theorem 4. The set of algebraic integers sitting in M together with their conjugates coincides with the values of c for which $z = 0$ is preperiodic under iterations of $f_c(z)$.

The proof is as for Theorem 1, but using the sequence $f_c^{(n)}(0)$ in the complex plane of the variable c. We remark that the algebraic integers fall into two different classes (i) c is such that 0 is periodic for $f_c(z)$, then $f_c(z)$ has a super-stable fixed point and $c \in \overset{\circ}{M}$, (ii) c is such that 0 is strictly preperiodic for $f_c(z)$, then $c \in \partial M$ and c is called a Misiurewicz point [10].

Very likely this argument which uses a kind of mixed iteration can be extended to many other similar examples.

7. CONCLUSION

Our results display a wide class of fractal sets in which it is possible to solve the localisation problem of algebraic integers. It is in fact not surprising that the study of the transformation of algebraic integers under polynomial functions with integer coefficients gives rise to interesting results. However, we do not know how to extend this to rational fractions. Let us also notice that our results may also be useful for the point of view of dynamical systems, since it gives number theoretical characterizations of points with simple finite orbit. One could reasonably hope that such an observation apply to much more general cases than those treated here.

REFERENCES

[1]   P. Moussa, J. S. Geronimo, D. Bessis, Ensembles de Julia
      et propriétés de localisation des familles itérées
      d'entiers algébriques, C. R. Acad. Sci. Paris, 299, Ser.
      I (1984), 281-284.

[2]   M. F. Barnsley, D. Bessis, P. Moussa, The Diophantine
      moment problem and the Analytic Structures in the Acti-
      vity of the Ferromagnetic Ising Model, J. Math. Phys. 20,
      (1979), 535-546.

[3]   P. Moussa, Problème diophantien des moments et modèle
      d'Ising, Ann. Inst. Henri Poincaré 38 (1983), 309-347.

[4]   C. Pisot, Sur les fonctions arithmétiques à croissance
      exponentielle, C.R. Acad. Sci. Paris 222 (1946), 1027-
      1028.

[5]   F. Gramain, Fonctions entières arithmétiques, Séminaire
      Lelong-Skoda, Lecture Notes in Mathematics 694 (1978),
      96-125, Springer; Fonctions entières arithmétiques,
      Séminaires Delange-Pisot-Poitou, 19ème année, n°8 (1977-
      1978), 1-14, Paris.

[6]   R. Creighton Buck, Integral valued entire functions,
      Duke Math. J. 15 (1948), 879-891.

[7]   M. Fekete, Über die Verteilung der Wurzeln bei Gewissen
      Algebraische Gleichungen mit ganzzähligen Koeffizienten,
      Math. Z. 17 (1923), 228-249.

[8]   For a recent review see P. Blanchard: Complex analytic
      dynamics on the Riemann sphere, Bull. Amer. Soc. 11
      (1984), 85-141.

[9]   B. B. Mandelbrot, Fractal aspects of the iteration of
      $z \to \lambda z(1-z)$ for complex λ and z, Annals N.Y. Acad. Sci.
      357 (1980), 249-259; On the quadratic mapping $z \to z^2 - \mu$
      for complex μ and $z \cdot$ the fractal structure of its M set
      and scalaing, Physica 7D (1983), 224-239.

[10] A. Douady, J. H. Hubbard, Itération des polynômes
      quadratiques complexes, C.R. Acad. Sci. Paris 294, Ser.
      I (1982), 123-216; A. Douady, Systèmes dynamiques
      holomorphes, séminaire Bourbaki n°599, Astérisque 105-
      106 (1983), 39-63.

[11] E. Dobrowoloski, On a question of Lehmer and the number
      of irreducible factors of a polynomial, Acta Arith. 34
      (1979), 391-401.

[12] D. H. Lehmer, Factorisation of certain cyclotomic func-
      tions, Ann. Math. 34 (1933), 461-479; For a review, see
      M. Waldschmidt, Sur le produit des conjugués
      extérieurs au cercle unité d'un entier algébrique,
      L'Enseignement Mathématique 26, Ser. II (1980), 201-209.

[13] A. Douady, Private communication.

REAL SPACE RENORMALIZATION AND

JULIA SETS IN STATISTICAL MECHANICS

Bernard Derrida

Service de Physique Théorique
CEN - Saclay
91191 GIF-SUR-YVETTE cedex France

1. INTRODUCTION

The Ising model is defined in the following way. One considers on each site i of a lattice, a spin variable S_i which can take two possible values $S_i = \pm 1$. A system of N Ising spins has therefore 2^N different configurations $C(S_i = \pm 1, \ldots, S_n = \pm 1)$. By definition of the model, the energy $H(C)$ of a given configuration is given by

$$H(C) = H(\{S_i\}) = - \sum_{ij} JS_i S_j \tag{1}$$

where the sum in (1) extends over all pairs of nearest neighbours on the lattice. For a given model of statistical mechanics, one wants to compute the partition function Z which is given by

$$Z = \sum_C \exp\left[\frac{-H(C)}{T}\right] = \sum_{S_i = \pm 1} \cdots \sum_{S_N = \pm 1} \exp\left[\frac{-H(\{S_i\})}{T}\right] \tag{2}$$

where T is the temperature. The reason why it is important to know Z is that from the knowledge of Z one can calculate all the thermal properties (energy, specific heat, etc....) as a function of temperature.

The Potts model is a generalization of the Ising model. On each site of a lattice, there is a variable σ_i which can take q possible values

$$\sigma_i = 1, 2, \ldots, q \tag{3}$$

By definition of the model, the energy $H(C)$ of a configuration is

$$H(C) = H(\{\sigma_i\}) = -\Sigma J \delta_{\sigma_i, \sigma_j} \tag{4}$$

where $\delta_{\sigma_i, \sigma_j}$ is the Kronecker delta.

Again for the Potts model, the goal is to calculate the partition function Z as a function of the temperature T

$$Z = \sum_{\sigma_1=1}^{q} \ldots \sum_{\sigma_N=1}^{q} \exp[-\frac{1}{T} H(\{\sigma_i\})] \tag{5}$$

In statistical mechanics, one is always interested in the properties of large systems ($N \sim 10^{23}$). Therefore the partition function is a huge sum (of 2^N terms for the Ising model and q^N for the Potts model).

One does not know in general how to calculate the partition function Z of the Ising or of the Potts model when the spins are located on a regular lattice in dimension $d \geq 2$ (with the famous exception of the Onsager Solution of the 2d Ising model). Therefore one has to use approximate methods to study these systems. One of the most efficient is the renormalization group approach.

2. RENORMALIZATION GROUP

Let us call Z(N,T) the partition function of an Ising or of a Potts model on a lattice of N sites at temperature T. The main goal of the Renormalization Group is to try to find a map T' = R(T) and a function g(T) such that

$$Z(N,T) = e^{Ng(T)} Z(N/b^d, T')$$
$$T' = R(T)$$

(6)

b is the scaling factor (larger than 1) and R(T) is the renormalization transformation which corresponds to the change of scale b.

One sees that if one can find g(T) and R(T) such that the partition function Z(N,T) satisfies (6), then one can calculate Z(N,T) for arbitrarily large N: by iterating the renormalization transformation one can reduce as much as one wants the size of the system and when the size is small enough, the calculation of Z using formula (2) or (5) becomes possible.

Unfortunately, one does not know in general how to find R(T) and g(T) for models on Bravais lattices (square, cubic, hypercubic, etc...). One knows only approximations of R(T) and g(T) [1].

3. HIERARCHICAL LATTICES

It was realized recently [2] that one can construct some scale invariant lattices (called hierarchical lattices) on which one can find R(T) and g(T) such that (6) is exact. These lattices are constructed by an iterative rule. The example of the diamond hierarchical lattice is described in Figure 1.

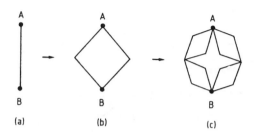

Fig. 1.   Construction of the diamond hierarchical lattice.

One starts at generation 1 with a lattice composed of 1 bond
and 2 sites (Fig. 1a). To go to the 2nd generation (Fig. 1b),
one replaces the bond of Fig. 1a by a set of four bonds. Then
one iterates the procedure. To go from generation n to generation n+1, one replaces each bond of the lattice at generation
n by a set of four bonds. It is easy to check that at generation n, the hierarchical lattice contains 4^{n-1} bonds and
$2(2 + 4^{n-1})/3$ sites.

By changing the rule of construction, one can construct as
many hierarchical lattices as one wants. The advantage of
these hierarchical lattices is that one can solve easily all
kinds of models of statistical mechanics.

Let us consider on the diamond hierarchical lattice of
Figure 1 a q state Potts model. On each site of the lattice,
there is a variable σ_i which can take q different values
($\sigma_i = 1,2,\ldots,q$) and the energy H of a configuration is given
by (4). It is convenient to introduce the variable x defined
by

$$x = e^{J/T} \tag{7}$$

If we call $Z_n(x)$ the partition function of the lattice constructed at the nth step, it is easy to relate Z_n and Z_{n-1}

$$Z_n(x) = Z_{n-1}(x')[A(x)]^{2 \cdot 4^{n-2}} \tag{8}$$

where $A(x)$ is given by

$$A(x) = 2x + q - 2 \tag{9}$$

and the renormalization transformation $R(x)$ which relates x to x' is given by

$$x' = R(x) = (\frac{x^2 + q - 1}{2x + q - 2})^2 \tag{10}$$

To obtain (8) for the lattice constructed after the nth generation, one just takes the trace over the sites which are connected to only two other sites. The remaining sites are on a hierarchical lattice of generation $n-1$ with an effective interaction x'. Since Z_1 is very easy to calculate

$$Z_1(x) = q(x + q - 1) \tag{11}$$

Formula (8) allows us to calculate all the Z_n.

If we define f_n as the free energy per bond in the lattice constructed after the nth generation then

$$f_n(x) = \frac{1}{4^{n-1}} \log Z_n(x) \tag{12}$$

and one sees from (8) that $f_n(x)$ obeys the following equation

$$f_n(x) = g(x) + \frac{1}{4} f_{n-1}(x') \tag{13}$$

where

$$g(x) = \frac{1}{2} \log(2x + q - 2) \tag{14}$$

In the thermodynamic limit (i.e., in the limit of an infinite system) f_n has a limit f

$$f = \lim_{n \to \infty} f_n \tag{15}$$

which obeys the following equation

$$f(x) = g(x) + \frac{1}{4} f(R(x)) \tag{16}$$

One can then calculate $f(x)$ for any x by using the convergent series

$$f(x) = \sum_{n=0}^{\infty} \frac{1}{4^n} g(R^n(x)) \tag{17}$$

4.   THE CRITICAL BEHAVIOUR OF f

The critical temperature is the temperature where the free energy is singular.   Therefore it is interesting to know where the function f given by (17) can be singular and what kind singularity one expects for f.

One expects (see the discussion below) that f is singular at all the unstable fixed points or the unstable periodic points of R.

The usual way of finding the critical behaviour of $f(x)$ at an unstable fixed point $x_c = R(x_c)$ is to assume that f has a power law singularity at x_c:

$$f_{sing}(x) = C|x - x_c|^{2-\alpha} \tag{18}$$

Then since $g(x)$ is analytic at the point x_c the singular parts coming from f in the two sides of (16) should be equal

$$C|x - x_c|^{2-\alpha} = \frac{1}{4} C|x - x_c|^{2-\alpha} [\frac{dR}{dx}(x_c)]^{2-\alpha} \tag{19}$$

This relates the exponent α to the slope of the renormalization transformation at the critical point x_c

$$\alpha = 2 - \log 4 / \log \left(\frac{dR}{dx} (x_c) \right) \tag{20}$$

So if f has a power law singularity (18) at x_c, then α is given by (20).

However, it is not at all obvious that f has any singularity at x_c nor that (18) gives the correct behaviour of the singular part of f near x_c.

Let us discuss briefly the more general following problem [4]. Given two functions $R(x)$ and $g(x)$ which are analytic at the fixed point x_c of the renormalization transformation R, is the function f solution of (16) singular x_c?

It is very easy to construct functions g for which f is analytic at x_c. If we choose an arbitrary function $Q(x)$ analytic at x_c, and if we consider for g a function g_m of the following form

$$g_m = Q(x) - \frac{1}{4^m} Q(R^m(x)) \tag{21}$$

then it is easy to see that the function f_m which solves (16) is

$$f_m = \sum_{p=0}^{m-1} \frac{1}{4^p} Q(R^p(x)) \tag{22}$$

Clearly f_m is a finite sum of analytic functions and therefore has no singularity at x_c.

It is interesting to notice that for any arbitrary function g (for which the solution of (16) has a singularity at x_c), one can choose $g_m(x) = g(x) - \frac{1}{4^m} g(R^m(x))$ which is very close to g and for which the function f_m has no longer any singularity.

Remark. Even if f has a singularity at x_c as it indeed does when $R(x)$ and $g(x)$ are given by (10) and (14) the critical behaviour near x_c is more complicated than the pure power law (18). Because of the discrete nature of the renormalization transformation, the power law is modulated [4,5]

$$f_{sing}(x) \sim |x - x_c|^{2-\alpha} h\left(\frac{\log|x - x_c|}{\log \frac{dR}{dx}(x_c)}\right) \tag{23}$$

where h is a periodic function of period 1 ($h(z + 1) = h(z)$). This oscillatory behaviour can be observed numerically by calculating f from (17). There exist also several approximations which give them. However, one does not know in general the exact expression of h in terms of R and g.

5. ZEROS OF THE PARTITION FUNCTION [6] - JULIA SET

An interesting question is to know the analyticity domain of the free energy $f(x)$ given by (17) or the limiting set of the zeros of the partition function $Z_n(x)$ in the thermodynamic limit. We shall see that the limiting set of zeros is the Julia set of the renormalization transformation $R(x)$.

First we can remark that the partition function $Z_n(x)$ is a polynomial of degree 4^{n-1} in the variable x. Therefore $Z_n(x)$ has 4^{n-1} zeros. Secondly, the recursion relation (8) between $Z_n(x)$ and $Z_{n-1}(x')$ gives a relation between the zeros of Z_{n-1} and Z_n. It is easy to see [6] that if $\tilde{x}_1$ is a zero of $Z_{n-1}(x)$, then its four preimages (i.e., the four solutions of $R(x) = \tilde{x}_1$) are zeros of $Z_n(x)$. Since one knows easily the zero of $Z_1(x)$ see (11), then the set of zeros of $Z_n(x)$ is just $R^{-(n-1)}(1-q)$ where $R^{-(n-1)}(x)$ denotes the 4^{n-1} preimages of x

by the renormalization transformation. In the limit $n \to \infty$
these zeros accumulate on the Julia set [7,8] of the map $R(x)$
since the Julia set is the accumulation set of the preimages
of any point. In Figure 2 one sees the Julia set (and therefore the set of zeros of Z_n) for $q = 2$ and $q = 4$.

Remark. If one considers more complicated hierarchical
lattices with competing interactions [4], the Julia set may
have a lot of points on the real axis. All these points are
possible critical temperatures of the physical model.

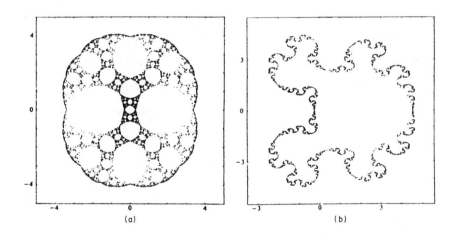

(a)   (b)

Fig. 2. Julia sets of the map $x' = R(x) = (\dfrac{x^2 + q - 1}{2x + q - 2})^2$
for $q = 2$ (figure a) and $q = 4$ (figure b).

6.   DISORDER ON HIERARCHICAL LATTICES

An important question in statistical mechanics is to know how the properties of a system are changed by the presence impurities.   One way of studying the effect of impurities is to consider that the energies of bonds are random variables

$$H = - \Sigma J_{ij} \delta_{\sigma_i \sigma_j} \tag{24}$$

i.e., the J_{ij} are not all equal but are distributed according to a given probability distribution.

Since it is easy to solve pure models on hierarchical lattices, it is interesting to know what can be said about models with random interactions.   Let us consider again a q state Potts model on the diamond hierarchical lattice [9,10] but now the nearest neighbour interactions x are randomly distributed according to a given probability distribution $P_0(x)$. Under one step of the renormalization transformation, a set of 4 different bonds x_1, x_2, x_3, x_4 becomes one bond

$$x = (\frac{x_1 x_2 + q - 1}{x_1 + x_2 + q - 2})(\frac{x_3 x_4 + q - 1}{x_3 + x_4 + q - 2}) = F(x_1, x_2, x_3, x_4) \tag{25}$$

Since the 4 x_i are random variables, x is also a random variable whose distribution $P_1(x)$ can be known if one knows $P_0(x)$ and the transformation F.   If we call $P_n(x)$ the probability distribution of a bond obtained after the nth step, then $P_n(x)$ can be calculated by the following recursion

$$P_n(x) = \int\int\int\int \prod_{i=1}^{4} P_{n-1}(x_i) dx_i \delta(x - F(x_1, x_2, x_3, x_4)) \tag{26}$$

So now instead of looking at the trajectory of a single point x, we have to follow the evolution of the probability

distribution $P_n(x)$.   Then the average free energy per bond is
given by

$$f(\{P_0\}) = \sum_{n=0}^{\infty} \frac{1}{4^n} \int\int dx_1 dx_2 P_n(x_1) P_n(x_2) \frac{1}{2} \log(x_1 + x_2 + q - 2)$$

(27)

The problem now is to find how the analytic properties of the
free energy depend on the distribution P_0.   This is not an
easy problem even if the distribution P_0 is very simple (for
example the sum of 2 delta functions).   The problem is that
when one iterates (26) the distribution P_n becomes more and
more complicated.

For example one can take for $P_0(x)$

$$P_0(x) = p\delta(x - y_1) + (1 - p)(x - y_2)$$

(28)

with $R^n(y_1) \to \infty$ and $R^n(y_2) \to 1$ as $n \to \infty$.   One knows very little
on the analytic properties of f as a function of p.

With Elizabeth Gardner [10], we tried to see how the
singular properties of f could be changed when the distribution
$P_0(x)$ is very narrow.   We obtained the following results valid
only for very narrow $P_0(x)$.

For $q < 4 + 2\sqrt{2}$, the exponents α_p and α_r of the pure and
random systems are equal.

For $q = 4 + 2\sqrt{2}$, the pure system has a specific heat which
diverges as $\log|x - x_c|$ whereas the random system has a speci-
fic heat singularity of the form $[\log|x - x_c|]^{(1-2\sqrt{2})/7}$.

For $q > 4 + 2\sqrt{2}$, α_p and α_r are different and for $\alpha_p \ll 1$,
one has $\alpha_r = \frac{1 - 2\sqrt{2}}{7} \alpha_p + O(\alpha_p^2)$.   (The exponent α is by definition related to the singular part of the free energy
$f \sim |x - x_c|^{2-\alpha}$.   When it is positive, α is the exponent of

the power law divergence of the specific heat).

7. CONCLUSION

We have seen that on hierarchical lattices, one can write exact renormalization transformations. For Bravais lattices, one does not know in general exact real space renormalizations and most the approximations which have used so far are not likely to work for complex values of temperatures.

With Henrik Flyvbjerg [11], we have proposed a way of constructing a sequence of real space renormalizations for Bravais which can work for complex temperatures. To construct one of these real space renormalizations one needs to know the partition functions of 3 different finite lattices: $Z_{N_1}(T)$, $Z_{N_2}(T)$ and $Z_{N_3}(T)$ denote the partition functions of the lattices with N_1, N_2, or N_3 sites. Here we shall choose for simplicity

$$N_2 = N_1/b^d \; ; \quad N_3 = N_2/b^d = N_1/b^{2d}$$

(For example in dimension 2 on a square lattice one can choose $b = 2$, $N_1 = 16$, $N_2 = 4$, $N_3 = 1$.)

The real space renormalization that we wrote was the following

$$\frac{Z_{N_1}(T)}{[Z_{N_2}(T)]^{b^d}} = \frac{Z_{N_2}(T')}{[Z_{N_3}(T')]^{b^d}} \tag{29}$$

When N_1, N_2, and N_3 increase, the values of the critical temperature and of the critical exponents are improved (we checked that for the 2d Ising model).

Since $Z_N(T)$ is a polynomial of the variable $\tanh(\frac{J}{T})$, the

relation (28) contains only polynomials. Thus, the map T → T'
can be defined for complex temperatures.

One sees that T' is an implicit function of T. So one has
to consider the Julia sets of implicit transformations. In
[11] we have drawn the Julia sets of the transformation (29)
for the 2d Ising model. They look rather similar to what is
expected for the set of zeros of the partition function.

Lastly, I would like to mention that Julia sets appear in
other physical problems like the spectral properties of lattices
or excitations of Magnetic Systems [12,13].

ACKNOWLEDGEMENTS

I would like to thank L. De Seze, J. P. Eckmann, A. Erzan,
H. Flyvbjerg, E. Gardner, C. Itzykson, J. M. Luck with whom
the different aspects presented here were developed.

REFERENCES

[1] Burkhardt, T. W. and Van Leeuwen, J. M. J., Real Space
     Renormalization, Vol. 30, 1982, Springer Verlag Berlin
     Heidelberg, New York.

[2] Berker, A. N. and Ostlund S., J. Phys. C12 (1979), 4961.

[3] Kaufman, M. and Griffiths, R. B., Phys. Rev. B24 (1981),
     496; Kaufman, M. and Griffiths, R. B., J. Phys. A15
     (1982), L239; Kaufman, M. and Griffiths, R. B., Phys.
     Rev. B30 (1984), 244; and references therein. Griffiths,
     R. B. and Kaufman, M., Phys. Rev. B26 (1982), 5022;
     Melrose, J. R., J. Phys. A16 (1983), 1041; Melrose, J. R.,
     J. Phys. A16 (1983), 3077.

[4] Derrida, B., Eckmann, J. P., Erzan, A., J. Phys. A16
     (1983), 893.

[5] Derrida, B., Itzykson, C., and Luck, J. M., Comm. Math.
     Phys. 94 (1984), 115.

[6] Derrida, B., De Seze, L., and Itzykson, C., J. Stat.
     Phys. 33 (1983), 559.

[7]   Julia, G., J. Math. Pures Appl. $\underline{4}$ (1918), 47.

[8]   For a recent review on Julia sets: Blanchard, P.,
       Bull. American Mathematical Society Vol. 11 (1984), 85-
       141.

[9]   Kinzel, W. and Domany, E., Phys. Rev. $\underline{B23}$ (1981), 3421.

[10]   Derrida, B. and Gardner, E., J. Phys. 17 (1984), 3223.
       Derrida, B., Dickinson, H. and Yeomans, J., J. Phys.
       $\underline{A18}$ (1985), L53.

[11]   Derrida, B. and Flyvbjerg, H., J. Phys. $\underline{A18}$ (1985),
       L313.

[12]   Southern, B. W. and Loly, P. D., J. Phys. $\underline{A18}$ (1985), 525.

[13]   Southern, B. W. and Loly, P. D., Solid State Sciences
       $\underline{54}$ (1984), 165.

REGULAR AND CHAOTIC CYCLING IN MODELS

FROM POPULATION AND ECOLOGICAL GENETICS

Marjorie A. Asmussen

Department of Mathematics
and
Department of Genetics
University of Georgia
Athens, Georgia

INTRODUCTION

Until recently, analysis of genetic models has typically ignored the possibility of regular and chaotic cycling in the genetic composition of a population. The reason for this neglect is twofold: (1) geneticists have not generally believed that such behavior occurs in natural populations except in cases involving fluctuating environmental conditions; and (2) no oscillatory behavior is possible in the classical models of population genetics. Now that the potential for limit cycles has been discovered within several genetic models, interest in this phenomenon is gradually increasing. Its importance has still by no means been widely accepted by geneticists, however.

The discussion which follows describes a few genetic contexts in which cycling in the genetic composition of a population has been reported. An interesting finding is that the nature of these cycles and the conditions under which they occur, are such that they are not apt to be detected experimentally. To set the stage for these more recent results, and

to introduce the necessary genetic terminology, I will begin by
describing briefly first the biological issues in population
genetics and secondly the key results from the classical selection model.

1. POPULATION GENETICS

Population genetics is concerned with the genetic composition of populations, including the forces affecting this composition and how it changes over time. Attention focuses on
the genetic <u>locus</u> (gene) or loci of an individual which determine a given genetic trait. A key finding is that many loci in
living organisms are <u>polymorphic</u>, meaning that they are represented by multiple forms (<u>alleles</u>). For example, in most
human populations there are three different alleles present at
the locus determining basic blood types: A, B, and O.

The single most important issue in population genetics is
why there is so much genetic variation in natural populations.
A major aim of mathematical genetic models is therefore to
determine under what conditions a given force or combination of
forces will maintain more than one allele at a locus. The
examples of cycling below all focus on the role of one particular force, namely natural selection, by which is meant the
differential reproduction and survival of the different
<u>genotypes</u> (genetic types) in a population. For an overview
of some of the many factors such as mating structure, migration,
genetic drift, etc. that can influence the genetic composition
of a population see [5] or [9].

2. CLASSICAL SELECTION MODEL

To appreciate the results from the more complicated models below it will be helpful to first review the classical model of natural selection introduced by R. A. Fisher in 1922. It is based on the following assumptions:

1. Two alleles, A_1 and A_2, at a single locus in a diploid organism (such as humans, where chromosomes occur in pairs). This means there are three possible genotypes in the population: A_1A_1, A_1A_2, and A_2A_2. The first and third of these, which carry two copies of the same allele, are called homozygotes and the second, which carries two different alleles, is called a heterozygote.

2. Random mating with respect to genotypes at the A locus.

3. Effectively infinite population size, so that the process is deterministic rather than stochastic.

4. The fitness of each genotype = the fraction of zygotes (newborn) which survive to reproduce, is constant over time.

5. No external source of genetic variation such as mutation or migration.

6. Discrete, nonoverlapping generations. The generation cycle is as follows

Mating pool (adults) gen. t	$\xrightarrow{\text{Random Mating}}$	Zygotes (Parents die)	$\xrightarrow{\text{Selection}}$	Mating pool (adults) gen. t+1

a. Recursion Equation

Under these assumptions, the recursion equation for p_t = frequency of allele A_1 in the mating pool in generation t, can be derived as follows:

Genotype:	A_1A_1	A_1A_2	A_2A_2
Relative freq. in zygotes in gen. t	p_t^2	$2p_t(1-p_t)$	$(1-p_t)^2$
Fitness	w_{11}	w_{12}	w_{22}
Relative freq. in mating pool in gen. t + 1	$p_t^2 \dfrac{w_{11}}{\bar{w}(p_t)}$	$2p_t(1-p_t)\dfrac{w_{12}}{\bar{w}(p_t)}$	$(1-p_t)^2\dfrac{w_{22}}{\bar{w}(p_t)}$

where $\bar{w}(p_t) = p_t^2 w_{11} + 2p_t(1-p_t)w_{12} + (1-p_t)^2 w_{22}$ is the mean

fitness in the population in generation t. The frequency of

A_1 in the mating pool in generation t+1 is then the probability

that an allele chosen at random from an adult in the mating

pool is A_1. By the law of total probability this is simply

$$\sum_{j=1}^{2} \text{Prob(allele is } A_1 | \text{allele is selected from } A_1A_j \text{ adult)} \cdot$$

Prob(adult in mating pool is A_1A_j), which becomes

$$p_{t+1} = p_t^2 w_{11}/\bar{w}(p_t) + (1/2)2p_t(1-p_t)w_{12}/\bar{w}(p_t)$$

or

$$p_{t+1} = p_t w_1(p_t)/\bar{w}(p_t) \qquad\qquad (1)$$

where $w_1(p_t) = p_t w_{11} + (1-p_t)w_{12}$.

b. Gene Frequency Dynamics

The gene frequency p_t in successive generations is thus

specified by the iterates of the transformation (1). Note that

under selective neutrality, where $w_{11} = w_{12} = w_{22}$, $p_t \equiv p_0$ for

all $t \geq 0$. Otherwise, (1) will generally be a highly nonlinear

equation. Its qualitative dynamics are nevertheless well

understood and can be predicted from the relative size of the

three genotypic fitnesses. (For details of this analysis, see

Chapter 4 of [5] and Table IX in Chapter 3 of [9].)

Two important features of the classical, constant selection model should be emphasized. First, genetic variation is preserved under (1) if and only if heterozygotes have a higher fitness than the two homozygotes (i.e., $w_{12} > w_{11}, w_{22}$, a condition known as <u>overdominance</u>). In all other cases the allele at a selective disadvantage is lost. Second, the gene frequency under (1) always either monotonically increases or monotonically decreases to an equilibrium value. <u>There are therefore never any oscillations in the gene frequency under the classical model</u>. With these facts in mind, let us now turn to some more recent models which produce much more complicated dynamical behavior.

3. ECOLOGICAL GENETICS AND DENSITY REGULATED SELECTION

<u>Most of the following discussion will focus on the behavior under density regulated selection</u>, in which the genotypic fitnesses are functions of the changing numbers of individuals in the population. The first models of this sort were introduced in the early 1970's as a consequence of experimental studies in ecological genetics which demonstrated conclusively that population size can dramatically influence the operation of natural selection.

a. Recursion Equations for Density Regulated Selection

The assumptions of the basic, density regulated model are the same as those given above for the classical model, with the exception that the fitness of each genotype $A_i A_j$ in generation t satisfies $w_{ij} = w_{ij}(N_t)$, where N_t is the total number

of individuals in the mating pool in generation t. There are
now two population variables:

> Genetic Component: the gene frequency p_t

> Ecological Component: the population size N_t.

The recursion system is based on the generation cycle:

Genotype:	A_1A_1	A_1A_2	A_2A_2
Numbers Before Selection	$p_t^2 N_t$	$2p_t(1-p_t)N_t$	$(1-p_t)^2 N_t$
Fitness	$w_{11}(N_t)$	$w_{12}(N_t)$	$w_{22}(N_t)$
Numbers After Selection	$p_t^2 N_t w_{11}(N_t)$	$2p_t(1-p_t)N_t w_{12}(N_t)$	$(1-p_t)^2 N_t w_{22}(N_t)$

In generation t+1 we therefore have

$$p_{t+1} = p_t w_1(p_t,N_t)/\bar{w}(p_t,N_t) \tag{2a}$$

$$N_{t+1} = N_t \bar{w}(p_t,N_t) \tag{2b}$$

where $w_1(p_t,N_t) = p_t w_{11}(N_t) + (1-p_t)w_{12}(N_t)$ and $\bar{w}(p_t,N_t) = p_t^2 w_{11}(N_t) + 2p_t(1-p_t)w_{12}(N_t) + (1-p_t)^2 w_{22}(N_t)$.

Several features are immediately evident from (2). First,
the equation for p_t is analogous to the classical version (1),
except that here $w_{ij} = w_{ij}(N_t)$. This is a nontrivial difference, however, because since the relative fitnesses vary
with N, their relative order may also change along a trajectory.
Thus, it is no longer possible to easily predict the trajectory
of gene frequency change. Indeed, under (2) changes in gene
frequency and population size are interdependent and hence
must be studied jointly.

b. Equilibrium Analysis

Due to the dynamical complexity of (2) attention has
naturally focused on the existence and local stability of joint
equilibria. Formally, the point $(\hat{p}, \hat{N})$ is a joint equilibrium
of (2) if and only if $(p_t, N_t) = (\hat{p}, \hat{N})$ implies $(p_{t+1}, N_{t+1}) =$
$(\hat{p}, \hat{N})$. Such a state is locally stable if $(p_t, N_t) \rightarrow (\hat{p}, \hat{N})$
whenever (p_0, N_0) is initially near $(\hat{p}, \hat{N})$. The mathematical
criterion for local stability of $(\hat{p}, \hat{N})$ is that both eigenvalues
of the matrix

$$
\begin{pmatrix}
\dfrac{\partial p_{t+1}}{\partial p_t}(\hat{p}, \hat{N}) & \dfrac{\partial p_{t+1}}{\partial N_t}(\hat{p}, \hat{N}) \\[3mm]
\dfrac{\partial N_{t+1}}{\partial p_t}(\hat{p}, \hat{N}) & \dfrac{\partial N_{t+1}}{\partial N_t}(\hat{p}, \hat{N})
\end{pmatrix}
\tag{3}
$$

have magnitude less than 1.

The density regulated model (2) has the fortuitous property that the product of the cross diagonal terms of (3) is
always 0 at a joint equilibrium. Consequently, the local stability eigenvalues are given explicitly by the two diagonal elements. Moreover, these have natural biological interpretations:

$$
\lambda_1 = \frac{\partial p_{t+1}}{\partial p_t}(\hat{p}, \hat{N}) = \text{the } \underline{\text{genetic component}} \text{ (how changes in}
$$
gene frequency near a joint equilibrium
depend on p alone)

$$
\lambda_2 = \frac{\partial N_{t+1}}{\partial N_t}(\hat{p}, \hat{N}) = \text{the } \underline{\text{ecological component}} \text{ (how changes}
$$
in population size near a joint equi-
librium depend on N alone).

The main point of interest here is that the ecological component of the local stability criterion always reduces to the
requirement that

$$-2 < \hat{N} \partial \bar{w}(\hat{p}, \hat{N}) / \partial N < 0. \tag{4}$$

For (4) to hold, increases in population size above the equilibrium level must decrease the mean fitness in the population,
but not too abruptly. As a result, it is possible to have no
locally stable equilibria under (2) if the population size
regulating mechanism reacts too sharply to perturbations of
the population size near joint equilibria.

c. Exponential Fitnesses

The consequences of such a situation were explored by
Asmussen [4]. The highlights of that investigation are given
below for the particular fitness form

$$w_{ij}(N) = \exp[r_{ij}(1-N/K_{ij})] \qquad i,j = 1,2. \tag{5}$$

These fitnesses are one in the class of fitnesses which are
nonnegative, continuously differentiable, decreasing functions
of the population size N. (The latter property is motivated by
the experimental observation that high densities are usually
detrimental due to increased competition for finite resources
such as food, nesting sites, etc.)

This class of fitnesses is further characterized by the
condition that $\lim\limits_{N\to\infty} w_{ij}(N) < 1 < w_{ij}(0)$. Associated with each
genotype are two ecological parameters: (1) a birth rate or
intrinsic rate of increase, r_{ij}, which reflects the fitness
at low population sizes; and (2) a carrying capacity, K_{ij},
which is the unique population size at which the fitness of
$A_i A_j$ is 1 (i.e., $w_{ij}^{-1}(1) = K_{ij}$).

Fitnesses in this class have up to three joint equilibria
with $\hat{N} > 0$. There always exist exactly two fixation equilibria:

$(\hat{p},\hat{N}) = (0,K_{22})$ and $(1,K_{11})$. There will also exist an <u>interior</u> equilibrium (with $0 < \hat{p} < 1$ and $\hat{N} > 0$) if and only if hetero- zygotes have a lower $(K_{12} < K_{11},K_{22})$ or higher carrying capa- city $(K_{12} > K_{11},K_{22})$ than the two homozygotes. For the exponential fitnesses in (5), the local stability conditions are:

	Genetic	+	Ecological
<u>Fixation</u> for A_i	$K_{12} < K_{ii}$	and	$0 < r_{ii} < 2$
<u>Interior</u> Equilibrium	$K_{12} > K_{11},K_{22}$	and	$-2 < \hat{N}\partial\bar{w}(\hat{p},\hat{N})/\partial N < 0$

Note that for these fitnesses, the genetic components of the local stability criteria depend only on the relative order of the three genotypic carrying capacity parameters. As long as not all three carrying capacities are identical, there will always be at least one joint equilibrium at which the genetic half of the stability criterion is met. <u>In this sense, the genetic system is inherently stable.</u> If the birth rate para- meters become too large, however, the population size regu- lating mechanism becomes unstable and there will be no locally stable, isolated equilibrium states.

What happens in such cases is suggested by the dynamics of the analogous, purely ecological model of population growth

$$N_{t+1} = N_t \exp[r(1-N_t/K)] \tag{6}$$

where r and K are the birth rate and carying capacity of a genetically homogeneous population. $\hat{N} = K$ is a globally stable equilibrium for (6) as long as $0 < r < 2$. As r increases above 2, however, N_t no longer converges to a single equili- brium value. Instead, as originally reported by May [12], N_t first converges to a limit cycle between two different

population sizes ($2 < r < 2.526$), then to a limit cycle between
four different population sizes ($2.526 < r < 2.656$), then to
an eight point limit cycle ($2.656 < r < 2.685$) and so on. Once
$r > 2.692$ the limit cycles of period 2^n become unstable and a
chaotic regime arises in which there are an uncountable number
of initial population sizes for which N_t does not settle down
into any finite cycle. For $r > 3.102$, there are cycles of
every possible integral period.

These findings have a direct bearing on the joint system
(2), because if a population becomes <u>fixed</u> for the allele A_i
(i.e., $p_t \to 0$ or 1), the subsequent dynamics are governed by
changes in N_t alone, namely by (6) with $r = r_{ii}$ and $K = K_{ii}$.
Fixation limit cycles therefore exist in the two dimensional system (2), with the gene frequency constant at 0 or 1
and the population size oscillating in the pattern found by
May. The critical question in this context is when will a
genetically variable population (i.e., one with $0 < p_0 < 1$)
converge to such a cycle. In other words, when will a fixation limit cycle be locally stable in the joint system?

Mathematically, a limit cycle between the $d = 2^n$ points

$$(\hat{p}_1, \hat{N}_1) \to (\hat{p}_2, \hat{N}_2) \to \dots \to (\hat{p}_d, \hat{N}_d) \tag{7}$$

is locally stable if and only if both eigenvalues of the matrix

$$\begin{pmatrix} \dfrac{\partial p_{t+d}}{\partial p_t}(\hat{p}_i, \hat{N}_i) & \dfrac{\partial p_{t+d}}{\partial N_t}(\hat{p}_i, \hat{N}_i) \\ \dfrac{\partial N_{t+d}}{\partial p_t}(\hat{p}_i, \hat{N}_i) & \dfrac{\partial N_{t+d}}{\partial N_t}(\hat{p}_i, \hat{N}_i) \end{pmatrix}$$

have magnitude less than 1, where p_{t+d} and N_{t+d} are formed by
iterating (2) over d generations. For the case of fixation

limit cycles, where each $\hat{p}_i = 0$ or each $\hat{p}_i = 1$ in (7), the
local stability criterion again splits into two components [4].
With fixation for A_i, the genetic half requires $K_{12} < K_{ii}$, as
at the isolated fixation equilibria, $(0, K_{22})$ and $(1, K_{11})$. The
ecological component is precisely that for the 1 dimensional
system (6) with $r = r_{ii}$ and $K = K_{ii}$. Thus, there will be a
locally stable fixation cycle for allele A_i of period 2^n, for
some positive integer n, as long as $K_{12} < K_{ii}$ and $2 < r_{ii} <$
2.692. The larger r_{ii} becomes in this range, the longer the
period of the cycle. The exact period and population sizes
along the cycle are those predicted from (6).

Further insight into the dynamical behavior of the joint
system is provided by numerically iterating (2) for various
initial conditions and parameter values. Some of the
interesting behavior which ensues is shown in Tables 1 and 2,
which are adapted from similar tables presented in [4].
Table 1 illustrates a case where there are no locally stable
equilibria. The period two fixation cycle, $(\hat{p}_1, \hat{N}_1) =$
$(0, 157.8) \rightleftarrows (\hat{p}_2, \hat{N}_2) = (0, 52.2)$, however, is always locally
stable, by the criteria given above. For $r_{12} \leq 2.3$, the population converged to this fixation cycle from all initial points
chosen. As r_{12} is increased above 2.3, (empirically) stable
interior limit cycles appear, with joint oscillations in p_t
and N_t. The period of these joint cycles increases from 2 to
4 to 8 etc. For still higher values of r_{12}, p_t and N_t can
both exhibit apparently chaotic cycles in the interior, with
no pattern evident after 2,000 generations. Whether the population converges to the stable fixation cycle or an interior
cycle depends on the initial gene frequency and population
size.

Consider next the examples in Table 2. For each parameter
set shown, the fixation equilibrium $(\hat{p}, \hat{N}) = (0, K_{22})$, is locally
stable. Nonetheless, as r_{12} is increased, interior limit
cycles appear of period 2, 4, 8, etc. These are again empirically locally stable in that they appear to have significant
domains of attraction. For still larger values of r_{12} some
trajectories exhibit apparently chaotic oscillations in the
interior despite the presence of a locally stable equilibrium.
Comparing the three rightmost columns of Table 2 reveals
another interesting point: increasing K_{22} and hence the genetic
stability of the fixation equilibrium, $(0, K_{22})$, delays the
appearance of stable and chaotic interior cycles. In this
sense, the genetic system has a stabilizing effect upon the
joint system.

Several important points are evident from these examples.
First, the biological conclusions from a model can be dramatically different when the potential for limit cycles is taken
into account. Tables 1 and 2 show, for instance, that contrary to the conclusion based on isolated equilibria, heterozygote advantage in carrying capacity is not necessary to maintain genetic variation. Second, interior limit cycles can be
simultaneously stable with fixation equilibria or fixation
cycles. Thus, unlike the classical model, initial conditions
can determine whether or not genetic variation is ultimately
lost.

Third, the fact that regular and even apparently chaotic
limit cycles can occur despite the existence of locally stable
fixation equilibria means that a lack of stable equilibria is
by no means necessary for the existence of limit cycles. This

Table 1. Limiting behavior of (p_t, N_t) for exponential fitnesses as a function of r_{12} when $r_{11} = 1.0$, $r_{22} = 2.2$, $K_{11} = 95$, $K_{12} = 100$, and $K_{22} = 105$. A limit cycle is denoted by a sequence of points $(\hat{p}_i, \hat{N}_i)$ connected by arrows, as in (7)

$r_{12} \leq 2.3$ (0, 157.8), (0, 52.2)

$r_{12} = 3.0$ (0, 157.8), (0, 52.2)   or*   (.51, 165.5), (.60, 43.2)

$r_{12} = 4.1$ (0, 157.8), (0, 52.2)   or*   (.78, 31.5), (.57, 221.4), (.83, 23.3), (.58, 190.0)

$r_{12} = 5.0$ (0, 157.8), (0, 52.2)   or*   "Chaos" in the interior

*(p_t, N_t) either converged to the fixation cycle or to the initial cycle shown, depending on the initial state (p_0, N_0).

Table 2. Limiting behavior of (p_t, N_t) for exponential fitnesses as a function of r_{12} for various values of K_{22}, where $r_{11} = r_{22} = 1.0$, $K_{11} = 95$, and $K_{12} = 100$. A limit cycle is denoted by a sequence of points connected by arrows, as in (7)

	$K_{22} = 105$	$K_{22} = 125$	$K_{22} = 150$
$r_{12} \leq 3.5$	(0,105)	(0,125)	(0,150)
$r_{12} = 4.0$	(0,105) or* $\begin{pmatrix}(.48,\ 267.2)\\(.39,\ 25.7)\end{pmatrix}$	(0,125)	(0,150)
$r_{12} = 4.6$	(0,105) or* $\begin{pmatrix}(.38,\ 9.8)\\(.490,304.8)\\(.41,\ 19.9)\\(.489,405.1)\end{pmatrix}$	(0,125) or* $\begin{pmatrix}(.45,310.9)\\(.23,\ 27.9)\end{pmatrix}$	(0,150)
$r_{12} = 5.0$	(0,105) or* "Chaos" in the interior	(0,125) or* $\begin{pmatrix}(.21,\ 12.2)\\(.474,342.3)\\(.26,\ 22.3)\\(.472,443.7)\end{pmatrix}$	(0,150)

*(p_t, N_t) either converged to the fixation equilibrium $(0, K_{22})$ or to the interior limit cycle shown, depending on the initial state, (p_0, N_0).

in turn raises the question as to whether other systems also have the potential for regular and chaotic cycles which has gone undetected. Lastly, the discovery that the genetic system can exert a stabilizing influence on the joint system may explain why cycling is rarely found in natural populations. (For further discussion of density regulated selection, see [4].)

4. OSCILLATORY BEHAVIOR IN OTHER GENETIC SYSTEMS

The density regulated system above is, to my knowledge, the only documented case in the literature where the genetic composition of a population can exhibit chaotic oscillations. Cyclical behavior has now been reported in a few purely genetic systems, but it is of a different sort. The regular and chaotic cycling found in the density regulated model is associated with real eigenvalues less than -1. In the other examples that are described in the brief space remaining, oscillations arise as a result of complex eigenvalues with magnitude greater than 1. The period of these cycles is thus often long and nonintegral.

a. Frequency-Dependent Selection

The first purely genetic example is provided by models of frequency-dependent selection, in which the genotypic fitnesses depend upon the allelic frequencies in the population. These are motivated by abundant experimental evidence that the fitness of an individual can depend upon the frequency of its genotype as well as the kinds and frequencies of the other genotypes in the population.

The fact that such models can produce oscillations in the

genetic composition of the population was early recognized by

Wright [14] and Kimura [11]. This potential was more exten-

sively documented by Allard and Adams [3] and Cockerham and

Burrows [6] for a class of frequency dependent models in which

individuals asexually produce exact genetic copies of them-

selves. The model assumes that an individual of type i has

fitness w_{ij} when associated with an individual of type j.

Assuming only pairwise interactions, the <u>net fitness</u> of an

individual of type i in generation t is then given by $w_{i,t} = \sum_{j=1}^{n} p_{j,t} w_{ij}$, where $p_{j,t}$ is the frequency of type j in the

population in generation t. Note the difference in notation

from the previous models. The frequencies of the various types

are then governed by the recursion system

$$p_{i,t+1} = p_{i,t} w_{i,t} / \bar{w}_t \qquad \text{for } i = 1,2,\ldots,n \qquad (8)$$

where $\bar{w}_t = \sum_{j=1}^{n} p_{j,t} w_{j,t}$.

Allard and Adams presented several numerical examples of

low frequency limit cycles in the genetic composition of a

population under (8). One set (see Figure 5 in [3]) involved

complementary interaction matrices $[w_{ij}]$ in which individuals

of different types exert equal but opposite effects upon each

other, relative to interactions between like types (i.e., for

$i \neq j$, $w_{ij} = 1 - \alpha < w_{ii} = w_{jj} = 1 < w_{ji} = 1 + \alpha$). Cockerham

and Burrows [6] subsequently explained the cyclic behavior

mathematically on the following grounds. The only interior

equilibrium which preserves the same types in the population

is unstable due to complex eigenvalues of the form $\rho e^{i\theta}$, where

ρ slightly exceeds 1. As a result, the population should

slowly move away from the interior equilibrium with oscillations of frequency roughly $2\pi/\theta$, whose amplitude will increase only imperceptibly with time.

An interesting feature of Allard and Adam's examples is that for given total interaction effects, the amplitude of the resulting fluctuations decreases rapidly as the number of different types in the population increases. They concluded that the many genotypes found in real populations make wide oscillations in genotypic frequencies unlikely. Furthermore, due to the low accuracy in physical measurements, cycling will not be seen even though it may actually be there with very low amplitude. A second series of examples (see Figure 7 in [3]) suggests that an increase in the number of genotypes affecting fitness also has a stabilizing effect.

b. Constant Selection at Two Loci

More recently, Hastings [10] has reported the occurrence of stable Hopf cycles in the two locus analog of the classical constant selection model, which eventually disappear as a result of a "blue sky" bifurcation. As in the frequency-dependent case, the limit cycles are associated with complex eigenvalues with magnitude greater than 1 at an interior equilibrium. The period of the gene frequency oscillations was always greater than 100 generations and usually not an integer. Each cycle consists of a long period spent near the equilibrium point and a relatively short time away from equilibrium with correspondingly rapid changes in gene frequency. Thus, once again, the nature of the cycles makes them unlikely to be detected experimentally, even if they in fact occur in nature. Similar behavior has been analyzed extensively by Akin ([1] and [2]) within the analogous continuous time model.

c. Gene-Cytoplasm Interactions

The last and to my knowledge, the most recent example of
cycling in a genetic system is provided by Gregorius and Ross
[7]. They introduced a complicated, discrete time model in
which the fitness of each genotype at a single diallelic locus
depends on the type of cytoplasm in which the nucleus (containing the chromosomes) is found. They present some numerical examples with joint oscillations in the gene frequencies
and the relative frequencies of the different cytoplasm types
in the population. The precise nature of the cycles, including
the underlying eigenvalues and period, is not specified, however.

5. CONCLUSIONS

The above discussion shows that oscillations in the genetic
composition of a population are indeed possible. Limit cycles
can even occur under constant selective values if the number
of loci involved is increased from 1 to 2. It is interesting
to note in this regard that analyses based on ecological models
have indicated that one dimensional models are far more stable
than higher dimensional ones and that cycling arises more
readily in systems of higher dimensions (see e.g., [12] and
[13]).

With density and frequency dependent selection, however,
increasing the dimension genetically can have a stabilizing
effect. Furthermore, this finding suggests one reason why
cycling is not often found in natural populations. A second
reason is that the low amplitude and long period of the cycles
in the frequency dependent examples and the two locus constant

selection model are such that they would tend to go undetected experimentally even if they actually occurred in a population.

These results show that the lack of experimental evidence for cycling by no means proves that such behavior is absent or unimportant in genetic systems. Indeed, the density regulated model above shows that when the potential for limit cycles is taken into account, the conditions for the maintenance of genetic variation are dramatically altered. The possibility of oscillations in the genetic composition of a population is therefore a critical issue which should be considered when analyzing and interpreting genetic models.

REFERENCES

[1] Akin, E., "Cycling in Simple Genetic Systems," J. Math. Biol. 13: 305-324 (1982).

[2] Akin, E., Hopf Bifurcation in the Two Locus Genetic Model, Memoirs American Mathematical Society #284 (1983).

[3] Allard, R. W. and Adams, J. "The Role of Intergenotypic Interactions in Plant Breeding," Proc. XII Intern. Congr. Genetics Vol. 3: 349-370 (1969).

[4] Asmussen, M. A., "Regular and Chaotic Cycling in Models of Ecological Genetics," Theor. Pop. Biol. 16: 172-190 (1979).

[5] Cavalli-Sforza, L. L. and Bodmer, W. F., The Genetics of Human Populations, W. H. Freeman and Company (1971).

[6] Cockerham, C. C. and Burrows, P. M., "Populations of Interacting Autogenous Components," Amer. Natur. 105: 13-29 (1971).

[7] Gregorius, H.-R. and Ross, M. D., "Selection with Gene-Cytoplasm Interactions I. Maintenance of Cytoplasm Polymorphisms," Genetics 107: 165-178 (1984).

[8] Guckenheimer, J., Oster, G., and Ipaktchi, A., "The Dynamics of Density-Dependent Population Models," J. Math. Biol. 4: 101-147 (1977).

[9]   Hartl, D., _Principles of Population Genetics_. Sinauer Associates, Inc. (1980).

[10] Hastings, A., "Stable Cycling in Discrete-Time Genetic Models," _Proc. Natl. Acad. Sci._ USA 78: 7224-7225 (1981).

[11] Kimura, M., "On the Change of Population Fitness by Natural Selection," _Heredity 12_: 145-167 (1958).

[12] May, R. M., "Biological Populations with Nonoverlapping Generations: Stable Points, Stable Cycles, and Chaos," _Science 186_: 645-647 (1974).

[13] May, R. M. and Oster, G. F., "Bifurcations and Dynamic Complexity in Simple Ecological Models," _Amer. Natur._ 110: 573-599 (1976).

[14] Wright, S., "Adaptation and Selection," In G. L. Jepsen, et al., _Genetics, Paleontology, and Evolution_. Princeton Univ. Press, pp. 365-389 (1949)

ACKNOLWEDGMENTS

Work on this paper was supported in part by National Science Foundation grant DEB-82-00664. This article was prepared while the author was a visiting Research Associate in the Department of Genetics at the University of California, Davis. The use of their facilities during this period is greatly appreciated. The author is also grateful for receipt of a Sarah H. Moss Fellowship through the University of Georgia which made this sabbatical year possible. The author would also like to thank R. W. Allard and A. Hastings for stimulating discussions of their work on genetic cycling.

A BIFURCATION GAP FOR A SINGULARLY

PERTURBED DELAY EQUATION

John Mallet-Paret[1]

Division of Applied Mathematics
Brown University
Providence, Rhode Island

Roger D. Nussbaum[2]

Department of Mathematics
Rutgers University
New Brunswick, New Jersey

ABSTRACT

The singularly perturbed differential-delay equation

$$\varepsilon \dot{x}(t) = - x(t) + f(x(t - 1))\qquad(1)$$

is compared with the discrete dynamical system

$$x_n = f(x_{n-1})\qquad(2)$$

for some model problems. In general, the dynamical structures
(periodic orbits, their stability properties, bifurcation
points) of (2) are <u>not</u> preserved in (1), no matter how small ε
is taken. Thus (2) does not furnish an accurate approximation
to (1) for small ε.

[1]Partially supported by NSF Grant MCS-8201768.
[2]Partially supported by NSF Grant MCS-8201316.

1. DELAY EQUATIONS AND INTERVAL MAPS

This paper concerns the relation between the singularly perturbed differential-delay equation

$$\varepsilon \dot{x}(t) = - x(t) + f(x(t - 1)) \tag{1}$$

and the discrete dynamical system

$$x_n = f(x_{n-1}). \tag{2}$$

Here $\varepsilon > 0$ is a small parameter, $x \in R$ is a scalar, and $f: R \to R$ is a given nonlinearity. Equation (1) arises in mathematical models in various areas of science including optics, physiology, and biology; see [1,7,10,11,12,13,15,17,18] and also the references in [2,19,20,21]. Equation (2) is obtained from (1) by formally setting $\varepsilon = 0$. (Note that one could rescale the time in (1) and consider the equivalent equation

$$\dot{x}(t) = - x(t) + f(x(t - 1/\varepsilon))$$

with a large time delay.)

The task at hand is to understand the dynamical behaviour of solutions of (1). One hopes that a knowledge of the discrete system (2) would be useful here. Indeed, one might expect that the dynamic structures (such as equilibria, periodic orbits, and chaotic attractors) found in (2) should be preserved, together with their stability properties, in the perturbed system (1): thus a fairly complete understanding of (1) for small ε should emerge from our knowledge of (2). Unfortunately, this is not generally the case, as our results will indicate. Except for some elementary situations, the dynamical behavior of (1) is usually quite different from, and

much richer than that of (2). Moreover this difference persists no matter how small ε is: in a certain sense the dynamical structures of (1) do not limit continuously to those of
(2) as ε → 0. The results in this paper will help to make this
idea precise. Proofs are omitted here as they either have
appeared [21] or will appear elsewhere.

To begin, we describe an elementary situation in which
some features of the dynamics of (2) are preserved in (1).
Recall the formulation of the initial value problem for (1):
having chosen a continuous function φ ∈ C[-1,0] as the prescribed initial condition

$$x \mid [-1,0] = \phi \tag{3}$$

for x(t) on [-1,0], one integrates equation (1) forward in time
to obtain uniquely (at least for continuous f) a function x(t)
continuous on [-1,∞) and satisfying the differential equation
(1) on (0,∞). (In general the solution through φ does not
exist for t < -1.) The following result concerns the situation
in which the initial condition takes values in a positively
invariant interval.

Proposition. Suppose f is continuous and that f(I) ⊂ I for
some compact interval I. Let x(t) be the solution of the
initial value problem (1), (3) for some ε > 0 and some initial
condition φ which takes values in I:

φ(t) ∈ I   for all   t ∈ [-1,0].

Then the solution also lies in I for all forward time:

x(t) ∈ I   for all   t ∈ [-1,∞).

In addition

$$x(t) \rightarrow I_\infty \quad \text{as} \quad t \rightarrow \infty$$

where $I_\infty \subset I$ is the set

$$I_\infty = \bigcap_{n=1}^{\infty} f^n(I)$$

with f^n denoting the nth iterate of f.

Thus the behaviour of the solution $x(t)$ starting in I mimics the behaviour of the sequence x_n given by (2) when $x_0 \in I$: one has $x_n \in I$ and $x_n \rightarrow I_\infty$ as $n \rightarrow \infty$.

A special case of the above proposition concerns behaviour near an attracting fixed point.

Corollary. Suppose f is continuous, that $f(a) = a$ for some $a \in R$, and that $f'(a)$ exists and satisfies $|f'(a)| < 1$. Then there exists $\kappa > 0$ such that for any initial condition satisfying

$$|\phi(t) - a| \leq \kappa \quad \text{on} \quad [-1,0],$$

the corresponding solution satisfies

$$|x(t) - a| \leq \kappa \quad \text{on} \quad [-1,\infty)$$

and

$$\lim_{t \rightarrow \infty} x(t) = a.$$

The above results, which are easily proved, are in fact true for all $\varepsilon > 0$; the singular perturbation structure plays no role. In contrast, Theorem 1 below is much more delicate as it describes the fine structure of some solutions as $\varepsilon \rightarrow 0$.

Before stating this result we make several definitions. A
slowly oscillating solution x(t) of equation (1) is a solution
on $(-\infty, \infty)$ such that the distance between any two of its zeros
is greater than the delay 1; that is, whenever $x(q) = x(p) = 0$
and $q \neq p$, then

$$|q - p| > 1.$$

Slowly oscillating solutions (first defined by Kaplan and
Yorke [14]) often occur when f satisfies the negative feedback
condition

$$xf(x) < 0 \quad \text{if} \quad x \neq 0 \tag{4}$$

with an instability condition

$$f'(0) < -1$$

at the origin. It is easy to see that if condition (4) holds,
and f is continuous, then all zeros of a slowly oscillating
solution are simple: $\dot{x}(q) \neq 0$ whenever $x(q) = 0$. Now suppose
that f satisfies the negative feedback condition (4) and let
$N \geq 1$ be an integer. We define a P_N-solution of equation (1) to
be a slowly oscillating solution satisfying the following properties:

(1) there exist numbers $\ldots < q_{-1} < q_0 < q_1 < q_2 < \ldots$ such
    that $x(q_n) = 0$ and $(-1)^n x(t) > 0$ in (q_n, q_{n+1}), for each n;
    and

(2) $x(t) = (-1)^N x(t + q_N)$ for all t.

Thus a P_N-solution repeats, with a possible sign change, after
N zeros. Note that if N is odd then f must be an odd function,
at least throughout the range of x(t). It will often be

convenient to normalize $q_0 = 0$ by a time translation; in this
case note that $x(0) = 0$ and $\dot{x}(0) > 0$. In any case we shall
always identify any solution $x(t)$ with its translates $x(t + \theta)$,
regarding them as the same. Elsewhere [20,21,22,23] P_1-
solutions have been called S-solutions, and P_2-solutions have
been called slowly oscillating periodic solutions.

In Mallet-Paret and Nussbaum [21] the following result is
proved.

Theorem 1. Suppose f is continuous and that

$$f(a) = -b \quad \text{and} \quad f(-b) = a$$

for some numbers $-b < 0 < a$. In addition, suppose there exists
a compact interval I containing both a and $-b$, and that

$f(I) \subset I$,

$xf(x) < 0 \quad \text{if} \quad x \in I - \{0\}$,

$f'(0)$ exists and satisfies $f'(0) < -1$, and

f is nonincreasing on some neighborhood of $x = 0$.

Finally, suppose that the iterates $x_n = f^n(x_0)$ given by the
discrete system (2) satisfy

$$x_n \to \{a,-b\} \quad \text{as} \quad n \to \infty$$

whenever $x_0 \in I - \{0\}$.

Then there exists $\varepsilon_* > 0$ such that if $0 < \varepsilon < \varepsilon_*$ then equation (1) possesses a P_2-solution satisfying

$$x(t) \in I \quad \text{for all} \quad t; \tag{5}$$

if f is an odd function then $x(t)$ may be chosen to be a P_1-
solution. There exist numbers $0 < R_1 < R_2$ such that if $x(t)$

is any P_2-solution satisfying (5) and normalized so $q_0 = 0$, then the zeros q_1 and q_2 satisfy

$$1 + \varepsilon R_1 \le q_1 \le 1 + \varepsilon R_2 \quad \text{and}$$

$$2 + \varepsilon R_1 \le q_2 \le 2 + \varepsilon R_2.$$

Given $\delta > 0$, there exist $\gamma > 0$ and $K > 0$ such that if $\varepsilon \le \gamma$ and $x(t)$ is any such solution then

$$|x(t) - sqw(t)| \le \delta \quad \text{on} \quad [\varepsilon K, 1 - \varepsilon K] \cup [1 + \varepsilon K, 2 - \varepsilon K]$$

where sqw is the period-two square wave function defined by

$$sqw(t) = \begin{cases} a & \text{on } [0,1) \\ -b & \text{on } [1,2) \end{cases}$$

$$sqw(t) = sqw(t + 2) \quad \text{for all } t.$$

2. THREE QUESTIONS

Theorem 1 describes a very precise relation between certain periodic points of the map f and periodic solutions of the differential equation (1). In particular, $x(t)$ converges to $sqw(t)$ uniformly on compact subsets of R-Z as $\varepsilon \to 0$. Encouraged by this, one might pose some general questions. For simplicitly, we suppose henceforth that all functions f that we consider satisfy the negative feedback condition (4). Define a $\underline{P_N\text{-point}}$ of a map f to be a nonzero $a \in R$ such that $(-1)^N f^N(a) = a$, where we require f to be an odd function if the integer N is odd. A $\underline{P_N\text{-orbit}}$ means the (finite) set $\{f^n(a) \mid n \ge 0\}$ of iterates of a P_N-point a. If a is P_N-point then define the periodic step function sqw(t;a) by

$$\text{sqw}(t;a) = f^n(a) \quad \text{on } [n,n+1) \quad \text{for} \quad 0 \le n \le N - 1$$

$$\text{sqw}(t;a) = (-1)^N \text{sqw}(t + N;a) \quad \text{for all } t.$$

Consider now the following three questions.

   Question 1. If $a \in R$ is a P_N-point of f, does there exist, for small ε, a (suitably translated) P_N-solution x(t) of (1) which converges to sqw(t;a) uniformly on compact subsets of R–Z as $\varepsilon \to 0$?

   Question 2. If the solution x(t) of Question 1 exists, and if the non-degeneracy condition $(-1)^N (f^N)'(a) \ne 1$ holds, is x(t) the unique such solution if ε is sufficiently small?

   Question 3. Again with x(t) as in Question 1 and ε small, if $|(f^N)'(a)| < 1$ then is x(t) an asymptotically stable solu-tion of (1), that is, do the non-trivial characteristic Floquet multipliers α of x(t) all satisfy $|\alpha| < 1$?

   Theorem 1 gives an affirmative answer to the first question when N = 1 or 2, provided f satisfies the conditions of that theorem. But in general, the (perhaps surprising) answer to each of the three questions is NO. This is so even in the simplest case N = 1.

3. A BIFURCATION GAP

   Before providing the counterexamples which yield negative answers to our questions, we first interpret the significance of such counterexamples in bifurcation theory. Suppose the map f depends on a bifurcation parameter $\mu \in R$; let us write $f_\mu(x)$. For a large class of such maps (see [4,5,8,9]) there exists a <u>period-doubling cascade</u>, namely parameter values

$\mu_1 < \mu_2 < \mu_3 < \ldots \rightarrow \mu_\infty$ such that for $\mu \in (\mu_n, \mu_{n+1})$ the map

f_μ possesses an attracting P_{2^n}-orbit (we are assuming for each

μ the normalization $f_\mu(0) = 0$ along with the negative feedback

condition (4)). For an odd function this cascade typically

starts with an attracting P_1-orbit for μ in some initial inter-

val (μ_0, μ_1). As μ increases past μ_n, the $P_{2^{n-1}}$-orbit loses its

stability, and from it the P_{2^n}-orbit bifurcates.

If Questions 1, 2, and 3 were answered affirmatively, one

would expect more and more of the cascade to be seen in equa-

tion (1) as $\varepsilon \rightarrow 0$: given any m, one should see bifurcation

points $\mu_n(\varepsilon)$ near μ_n, and an attracting P_{2^n}-solution of equa-

tion (1) for $\mu \in (\mu_n(\varepsilon), \mu_{n+1}(\varepsilon))$, for small enough ε and

$0 \le n \le m$. However, as the answers to our questions are nega-

tive in general, one might instead expect the following alter-

nate scenario for some functions f: the cascade for equation

(1) is of bounded length, breaking down at some fixed m_*

<u>independent of</u> ε. The mechanism of such a breakdown could con-

ceivably occur as a result of new bifurcation points appearing.

For example, as shown in Figure 1 there could exist a fixed

$\kappa > 0$ and a new bifurcation point

$$\mu_*(\varepsilon) \in (\mu_0 + \kappa, \mu_1 - \kappa)$$

at which the P_1-solution of (1) loses its stability. This pre-

mature loss of stability at $\mu_*(\varepsilon)$ rather than at some $\mu_1(\varepsilon)$

near μ_1, and the gap $(\mu_*(\varepsilon), \mu_1)$ of uniform size at least κ,

has indeed been observed in some physical models [11]. A

phenomenon of this sort in which a bifurcation cascade is

truncated at a finite point has been termed a <u>bifurcation gap</u>

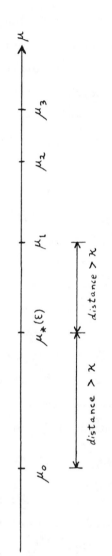

Fig. 1. A new bifurcation point $\mu_*(\varepsilon)$ a uniform distance κ away from μ_0 and μ_1.

[6].

We note that the bifurcation at such a point $\mu_*(\varepsilon)$ need not be a period doubling; as equation (1) describes an infinite dimensional system, it is conceivable that more complicated phenomena such as the bifurcation of an invariant torus could occur. Indeed, the existence of a rich structure of bifurcations and a transition to chaos for equation (1), all within the interval (μ_0, μ_1) is consistent with our results.

4. THE COUNTEREXAMPLES

Negative answers to Questions 1, 2, and 3, for the case $N = 1$, are furnished by the two-parameter family of non-linearities $f_{c,w}$ defined by

$$f_{c,w}(x) = \begin{cases} -w & \text{on } (0,c) \\ -1 & \text{on } (c,\infty) \end{cases}$$

$$f_{c,w}(-x) = -f_{c,w}(x) \qquad \text{for all } x,$$

where $c > 0$ and $w > 0$ are appropriately chosen; the values of $f_{c,w}$ at the discontinuities $x = 0, \pm c$ are immaterial and may be left undefined. (Although our counterexamples here are somewhat idealized -- the functions $f_{c,w}$ are discontinuous -- there is strong evidence that this feature is not essential.)

The following result gives a negative answer to Question 1. For the range of c and w considered in this theorem, the map $f_{c,w}$ has two P_1-orbits, namely, $\{1,-1\}$ and $\{w,-w\}$. (A third "ideal" P_1-orbit, $\{c,-c\}$, would appear if the jumps in the graph of $f_{c,w}$ were "filled in" with vertical lines.) We show that if $\frac{1}{2}(1 + w) \le c < 1$ then there does not exist a

P_1-solution of (1) corresponding to the orbit $\{1,-1\}$.

Theorem 2. Fix c and w to satisfy $0 < w < c < 1$. If $c \geq \frac{1}{2}(1 + w)$ then for all small ε equation (1) with $f = f_{c,w}$ has a unique P_1-solution. This solution (translated so $q_0 = 0$) satisfies

$$\lim_{\varepsilon \to 0} x(t) = sqw(t;w) \tag{6}$$

uniformly on compact subsets of R-Z. On the other hand, if $c < \frac{1}{2}(1 + w)$ then for all small ε there are exactly three P_1-solutions; two of them satisfy (6) while the other (suitably translated) satisfies

$$\lim_{\varepsilon \to 0} x(t) = sqw(t;1)$$

uniformly on compact subsets of R-Z.

The fact that the transition between the two types of behaviour in Theorem 2 occurs on a straight line $c = \frac{1}{2}(1 + w)$ in parameter space is unexpected; a priori there seems no reason why this should be so.

The next result gives negative answers to Questions 2 and 3. Note that in this theorem there is only one P_1-orbit namely $\{1,-1\}$, and that it attracts all initial conditions $x_0 \neq 0$, that is, $f^n(x_0) \to \{1,-1\}$. Thus (except for the fact that f is discontinuous) we are in the setting of Theorem 1.

Theorem 3. There exist sets $T,U \subset (0,1) \times (1,\infty)$ with non-empty interior, possessing the following properties. If $(c,w) \in T$ is fixed, then for all small ε equation (1) with $f = f_{c,w}$ has exactly three P_1-solutions. If $(c,w) \in U$ is fixed, then for all small ε equation (1) with $f = f_{c,w}$ has a

unique P_1-solution; this solution has a characteristic multiplier $\alpha > 1$ and hence is unstable.

Indeed, for any fixed c and w satisfying $0 < c < 1$ and
$w > c$ exactly one of the following holds with $f = f_{c,w}$:
(1) For all small ε equation (1) has a unique P_1-solution; or
(2) for all small ε equation (1) has exactly three P_1-solutions.
Furthermore, any P_1-solution of (1), for small ε, satisfies
the conclusions of Theorem 1 with $I = [-1,1] \cup [-w,w]$ and
$a = b = 1$.

Unlike the linear relation between c and w in Theorem 2,
the descriptions of sets T and U of Theorem 3 are rather complicated sets of polynomial inequalities. We omit the formulas
for these but do at least state, by way of example, that

$$(\tfrac{1}{5}, 6) \in \text{int } T \quad \text{and} \quad (\tfrac{1}{2}, 2) \in \text{int } U.$$

Figure 2 depicts the three P_1-solutions for $(c,w) = (\tfrac{1}{5}, 6)$,
for small ε. Observe that these solutions overshoot the values
± 1 near the jump points by an amount which <u>does not become</u>
<u>small</u> as $\varepsilon \to 0$; for example the range of the largest solution
shown in Figure 2 approaches the interval $[-\tfrac{11}{6}, \tfrac{11}{6}]$ and not
$[-1,1]$ for small ε. This non-uniform convergence of $x(t)$ to
the square wave sqw$(t;1)$ resembles the Gibbs phenomenon of
classical Fourier series, and holds quite generally when f is
not monotone in x; see [19,20,21]. Moreover, it is this feature of $x(t)$, and generally its behaviour near the jump points,
which seems to cause the instability described in Theorem 3.
One could think of the overshoot here as a "noise" super-
imposed on the square wave which destabilizes the solution,
thereby giving rise to a bifurcation gap.

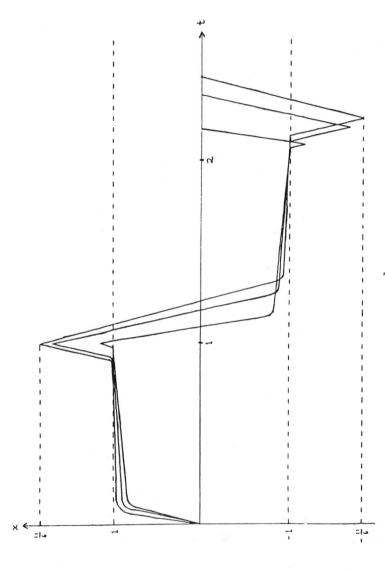

Fig. 2. The three P_1-solutions for $(c,w) = (\frac{1}{5}, 6)$ and $0 < \varepsilon << 1$. The sup norms of these solutions approach $1.1613\ldots$, $1.6877\ldots$, and $\frac{11}{6} = 1.8333\ldots$ as $\varepsilon \to 0$. The peaks which overshoot the values $x = \pm 1$ have width $O(\varepsilon)$.

5. A NEW DYNAMICAL SYSTEM

If the discrete system (2) does not accurately describe equation (1) for small ε, what then does? Can one find a new dynamical system, not depending on ε, which reflects at least some of the dynamics of (1) when ε is small? We believe the answer to this question is <u>YES</u>, and below we shall exhibit a map F, related to but different from f, as a candidate for such a system.

To understand the behaviour of solutions of (1) when ε is small, we study them near the jump points of the step function sqw(t;a). Consider a P_N-point a of the map f, and suppose there exists an associated P_N-solution x(t) of equation (1) as described in Question 1. To be definite assume a > 0 and $q_0 = 0$. One expects that as in Theorem 1, x(t) should be near $f^n(a)$ (say within δ) throughout the interval $[n + \varepsilon K, n + 1 - \varepsilon K]$ for $0 \le n \le N$ and for some $K = K(\delta) > 0$ and small enough ε. In addition, for such n the zeros q_n of x(t) should lie in the interval $[n, n + \varepsilon R]$ for some R > 0 independent of ε. (Recall that $q_{n+1} - q_n > 1$ hence $q_n > n$ for positive n.) Define quantities r_n by setting

$$q_{n+1} - q_n = 1 + \varepsilon r_n \tag{7}$$

and note that

$$0 < r_n \le R. \tag{8}$$

Define the functions $y_n(t)$ by stretching the time about the zero q_n:

$$y_n(t) = x(q_n + \varepsilon t).$$

Then from the differential equation (1), and from (7) one has

$$\dot{y}_n(t) = -y_n(t) + f(y_{n-1}(t + r_{n-1})) \tag{9}$$

for $1 \le n \le N$. Assuming a uniform boundedness condition
$x(t) \in I$ for all t where I is a fixed compact interval, and
using the bound (8), we may take limits of $y_n(t)$ and r_n as
$\epsilon \to 0$ (by possibly passing to subsequences $\epsilon_k \to 0$, $y_{nk}(t) \to$
$y_n(t)$, and $r_{nk} \to r_n$) to obtain functions $y_n(t)$ satisfying (9)
for all real t. These functions possess limiting values

$$y_n(-\infty) = f^{n-1}(a) \quad \text{and} \quad y_n(\infty) = f^n(a) \tag{10}$$

along with the (presumably strict) sign condition

$$(-1)^n t y_n(t) > 0 \quad \text{for all} \quad t \ne 0.$$

The periodicity of $x(t)$ implies that

$$y_0(t) = (-1)^N y_N(t) \quad \text{for all} \quad t.$$

Finally, integrating equation (9) and using the boundary
conditions (10) yield

$$y_n(t) = \int_{-\infty}^{t+r} e^{s-t-r} f(y_{n-1}(s)) ds \quad \text{where } r = r_{n-1}. \tag{11}$$

We now interpret formula (11) dynamically. Consider the
set

$$Y = \{y: R \to R \mid y(t) \text{ is continuous,}$$
$$y(\pm\infty) \text{ exist, are finite, and satisfy}$$
$$y(\infty) = f(y(-\infty)) \ne 0,$$
$$y(t) = 0 \text{ if and only if } t = 0, \text{ and}$$
$$y(t) \text{ changes sign at } t = 0\}$$

and define a map

$$F: Y \to Y$$

as follows. If $y \in Y$ then define $F(y) \in Y$ to be

$$F(y)(t) = \int_{-\infty}^{t+r} e^{s-t-r} f(y(s)) ds \tag{12}$$

where $r = r(y) \in R$ is the unique value such that $F(y)(0) = 0$.
(The negative feedback condition easily implies that a unique
r exists -- it is simply the horizontal translation of the graph
of $F(y)$ so that is passes through the origin -- and one sees
that $F(y) \in Y$.) Rewriting (11) as

$$y_n = F(y_{n-1})$$

we observe that y_0 is a P_N-point for the map F:

$$y_0 = (-1)^N y_N = (-1)^N F^N(y_0).$$

We have thus formally made a correspondence between
P_N-solutions of (1) and P_N-orbits of the map F. Indeed, the
estimates of Theorem 1 imply that in that case our derivation
of a P_N-orbit for F is rigorous. (In [21] this is presented
in a somewhat different style, where in place of the map F
a pair of so-called <u>transition layer equations</u>, equivalent to
the differential equations (9), is studied.) The above
analysis also explains the phenomena of Theorem 2 and 3:
calculations reveal that for $f = f_{c,w}$ there is a one-to-one
correspondence between P_1-solutions of (1) and P_1-orbits of F,
at least when all P_1-orbits are nondegenerate. In addition,
the stability properties of these orbits of F are preserved
in the corresponding solutions $x(t)$ of (1), for small ε.

We believe the correspondence between P_N-orbits of F and P_N-solutions of (1) holds in general. To be precise we make the following conjecture.

Conjecture. Questions 1, 2, and 3 have an affirmative answer if one replaces

$$
\begin{array}{lll}
f & \text{with} & F, \\
a \in R & \text{with} & y \in Y, \\
\text{sqw}(t;a) & \text{with} & \text{sqw}(t;y(\infty)), \\
(-1)^N(f^N)'(a) \neq 1 & \text{with} & 1 \notin \sigma((-1)^N DF^N(y)), \text{ and} \\
|(f^N)'(a)| < 1 & \text{with} & \rho(DF^N(y)) < 1,
\end{array}
$$

where $DF^N(y)$ denotes the derivative of F^N at y, and where $\sigma(L)$ and $\rho(L)$ denote the spectrum and the spectral radius of the linear operator L.

A number of functional analytic issues must be confronted before F can be properly studied as a dynamical system. The derivative DF(y) is as yet undefined, as the underlying space Y is not linear; indeed, we have not even endowed Y with a topology. Formally differentiating (12), keeping in mind the dependence $r = r(y)$, yields

$$
[DF(y)(z)](t) = \int_{-\infty}^{t+r} e^{s-t-r} f'(y(s)) z(s) ds
$$

$$
+ \frac{1}{f(y(r))} \left[\int_{-\infty}^{t+r} e^{s-t-r} f(y(s)) ds - f(y(t+r)) \right]
$$

$$
\times \int_{-\infty}^{r} e^{s-r} f'(y(s)) z(s) ds
$$

for $y \in Y$ and $z: R \to R$. The derivative of F^N can now be defined from this formula using the chain rule. A natural topology for Y is that of the Banach space

$$X = \{y:R \to R \,|\, y(t) \text{ is continuous, } y(\pm\infty) \text{ exist}$$
$$\text{and are finite, and } y(0) = 0\}$$

endowed with the supremum norm; with this the spectrum of
$DF^N(y)$ as an operator on X is defined.

Note that Y is not a closed subset of X; also Y has empty

interior in X. And although F is continuous on Y, it is not

in general completely continuous: the image of a bounded set

in Y need not be precompact in X (this can be seen if f is

linear or piecewise linear).

There is nevertheless hope for understanding the map F, as

is seen by studying the case of the step function nonlinearities

$f_{c,w}$. We denote $F = F_{c,w}$ in this case. If $0 < c < 1$ and

$w > c$ are fixed, then there is a compact attractor $\Gamma \subset Y$ for

$F_{c,w}$ which is homemorphic to a disjoint pair of intervals.

Considering the map $-F_{c,w}$ restricted to one of these intervals

(which we can do by symmetry) is equivalent to studying a

particular continuous map

$$G_{c,w}: [-1,1] \to [-1,1]$$

on a canonical interval $[-1,1]$. The map $G_{c,w}$ can be given by

an explicit though complicated formula ($G_{c,w}(\xi)$ is piecewise

a linear fractional transformation in ξ) which we omit; the

graph of $G_{c,w}$ for typical values of c and w is shown in

Figure 3. In contrast to the simple dynamics of $f_{c,w}$, which

have a unique globally attracting periodic orbit $\{1,-1\}$, the

map $G_{c,w}$ (and hence $F_{c,w}$) can have extremely complicated

dynamics. For example $G_{c,w}$ can possess a period three orbit

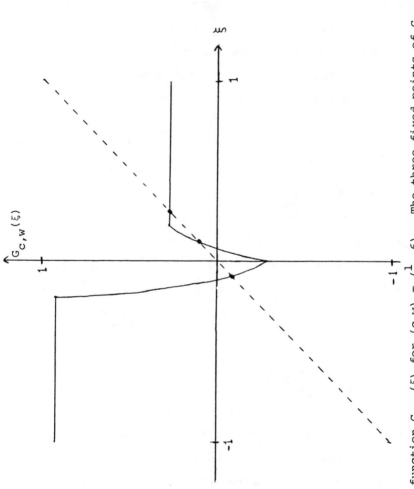

Fig. 3. The function $G_{c,w}(\xi)$ for $(c,w) = (\frac{1}{5}, 6)$. The three fixed points of $G_{c,w}$ correspond to the P_1-solutions of Figure 2.

and hence, by results of Sharkovskii [24] and Li and Yorke [16], has orbits of each period as well as uncountably many non-periodic orbits any two of which have different asymptotic behaviour.

In the following theorem a _total orbit_ of F means a bi-infinite sequence $\{y_n\}_{n=-\infty}^{\infty}$ in Y such that

$$y_n = F(y_{n-1}) \quad \text{for all} \quad n \in Z.$$

Also, $\| \ \|$ denotes the norm in X, that is, the supremum norm.

Theorem 4. There exists a set $C \subset (0,1) \times (1,\infty)$ with non-empty interior, such that for each $(c,w) \in C$ the map $F_{c,w}$ possesses the following properties. For each integer $N \geq 1$, $F_{c,w}$ has a P_N-orbit which is not a P_M-orbit for any $M < N$. In addition, there exists an uncountable set Σ of total orbits of $F_{c,w}$ such that for any two (not necessarily different) such orbits $\{y_n\}$ and $\{z_n\}$ and any integer k one has either

$$\limsup_{n \to \infty} \|y_n - z_{n+k}\| > 0 \quad \text{and} \tag{13}$$

$$\limsup_{n \to -\infty} \|y_n - z_{n+k}\| > 0, \tag{14}$$

or else one has $k = 0$ and $y_n = z_n$ for all $n \in Z$.

It follows easily from Theorem 4 that (13) and (14) also hold whenever $\{y_n\}$ is a total orbit in Σ and $\{z_n\}$ is any periodic orbit of $F_{c,w}$. One expects that each element of Σ should correspond to a slowly oscillating solution of (1) which is not periodic.

We state without proof that

$$(\tfrac{1}{5}, 6) \in \text{int } C.$$

Our final result concerns the parameter range for which
$f_{c,w}$ is monotone and possesses a unique P_1-orbit.   In such a
case $F_{c,w}$ in fact possesses a unique P_1-orbit which is a global
attractor.   The relation between monotonicity of a general nonlinearity f and uniqueness of P_1- and P_2-orbits of F is
explored by Chow, Lin, and Mallet-Paret [3].

Theorem 5.   Fix c and w to satisfy $0 < c < w \le 1$.   Then the
map $F_{c,w}$ possesses a unique P_1-orbit $\{y_*, -y_*\}$ in Y.   For any
$y \in Y$ one has

$$F_{c,w}^n(y) \to \{y_*, -y_*\} \text{ as } n \to \infty$$

in the topology inherited from X.

REFERENCES

[1]   S. P. Blythe, B. M. Nisbet, and W. S. C. Gurney,
      "Instability and complex dynamic behaviour in population
      models with long time delays," Theor. Pop. Biol. 2
      (1982), 147-176.

[2]   S.-N. Chow and J. Mallet-Paret, "Singularly perturbed
      delay-differential equations," Coupled Nonlinear
      Oscillators, ed. by J. Chandra and A. C. Scott, North
      Holland Math. Studies, Vol. 80 (1983), 7-12.

[3]   S.-N. Chow, J. Mallet-Paret, and X.-B. Lin, in prepara-
      tion.

[4]   P. Collet and J.-P. Eckmann, "Iterated Maps on the
      Interval as Dynamical Systems," Birkauser, 1980.

[5]   W. A. Coppel, "Maps on an interval," Institute for
      Mathematics and its Applications, Preprint No. 26, Univ.
      Minnesota, June, 1983.

[6]   J. P. Crutchfield and B. A. Huberman, "Fluctuations in
      the onset of chaos," Phys. Lett. 77A (1980), 407-410.

[7]   M. W. Derstine, H. M. Gibbs, F. A. Hopf, and D. L.
      Kaplan, "Alternate paths to chaos in optical bistability,"
      Phys. Rev. A 27 (1983), 3200-3208.

[8]   M. Feigenbaum, "Quantitative universality for a class
      of non-linear transformations," J. Statis. Phys. 19
      (1978), 25-52.

[9]   M. Feigenbaum, "The universal metric properties of non-
      linear transformations," J. Statist. Phys. 21 (1979),
      669-706.

[10] U. an der Heiden and M. C. Mackey, "The dynamics of pro-
      duction and destruction: analytic insight into complex
      behaviour," J. Math. Biol. 16 (1982), 75-101.

[11] F. A. Hopf, D. L. Kaplan, H. M. Gibbs, and R. L.
      Shoemaker, "Bifurcation to chaos in optical bistability,"
      Phys. Rev. A 25 (1982), 2172-2182.

[12] K. Ikeda, H. Daido, and O. Akimoto, "Optical turbulence:
      chaotic behaviour of transmitted light from a ring
      cavity," Phys. Rev. Lett. 45 (1980), 709-712.

[13] K. Ikeda, K. Kondo, and O. Akimoto, "Successive higher-
      harmonic bifurcations in systems with delayed feedback,"
      Phys. Rev. Lett. 49 (1982), 1467-4170.

[14] J. L. Kaplan and J. A. Yorke, "On the stability of a
      periodic solution of a differential delay equation,"
      SIAM J. Math. Anal. 6 (1975), 268-282.

[15] A. Lasota, "Ergodic problems in biology," Astérisque 50
      (1977), 239-250.

[16] T.-Y. Li and J. A. Yorke, "Period three implies chaos,"
      Amer. Math. Monthly 82 (1975), 985-992.

[17] M. C. Mackey and L. Glass, "Oscillation and chaos in
      physiological control systems," Science 197 (1977),
      287-289.

[18] M. C. Mackey and U. an der Heiden, "Dynamical diseases
      and bifurcations: understanding functional disorder in
      physiological systems," Funkt. Biol. Med. 1, 156 (1982),
      156-164.

[19] J. Mallet-Paret, "Morse decompositions and global con-
      tinuation of periodic solutions for singularly per-
      turbed delay equations," Systems of Nonlinear Partial
      Differential Equations, ed. by J. M. Ball, D. Reidel,
      Dordrecht, 1983.

[20] J. Mallet-Paret and R. D. Nussbaum, "Global continuation
      and complicated trajectories for periodic solutions of a
      differential-delay equation," Proceedings of Symposia in
      Pure Mathematics, American Mathematical Society, to
      appear.

[21] J. Mallet-Paret and R. D. Nussbaum, "Global continuation
      and asymptotic behaviour for periodic solutions of a
      differential-delay equation," submitted for publication.

[22]   D. Saupe, "Accelerated PL-continuation methods and
       periodic solutions of parameterized differential-delay
       equations," Ph.D. dissertation (in German), Univ. Bremen,
       1982.

[23]   D. Saupe, "Global bifurcation of periodic solutions of
       some autonomous differential delay equations,"
       Forschungschwerpunkt Dynamische Systeme, Report No. 71,
       Univ. Bremen, July, 1982.

[24]   A. N. Sharkovskii, "Coexistence of cycles of a continuous
       map of the line into itself " (in Russian), <u>Ukr. Mat.
       Zhur.</u> 16 (1964), 61-71.

TRAVELLING WAVES FOR FORCED FISHER'S EQUATION

Lawrence Turyn
Department of Mathematics and Statistics
Wright State University
Dayton, Ohio

1. INTRODUCTION

Fisher [3] and Kolmogorov, Petrovskii, and Piskunov [6] introduced Fisher's partial differential equation

$$u_t = u_{xx} + f(u) \tag{1.1}$$

to model population growth. When $f(u) = u(1 - u)$, these authors established the existence of <u>travelling wave</u> solutions $u(x,t) = U(x - ct)$ for $c \geq 2^{1/2}$, where $U(z)$ is positive for all z and has limits $U(\infty) = 0$, $U(-\infty) = 1$.

Fife [2] is a good reference for existence of travelling waves for (1.1) under hypothesis

$$f(0) = f(1) = 0 \text{ and } f \in C^2(\mathbb{R}, \mathbb{R}) \tag{H0}$$

and <u>one</u> of the hypotheses

$$f'(0) > 0 > f'(1) \text{ and } f \text{ positive on interval } (0,1) \tag{H1}$$

$$f'(0) < 0, \ f'(1) < 0 \text{ and } f \text{ has exactly one zero in } (0,1) \tag{H2}$$

The results are in two broad cases:

<u>Case 1</u>: With (H0) and (H1) there is a wave U_c for all

speeds $c \geq c^*(f)$ = minimal speed, and

Case 2: With (H0) and (H2) there is a unique speed c^* for which there is a wave.

In fact, for Case 1 there are two disjoint sub-cases:

Case 1(a): $c^*(f) = (2f'(0))^{1/2}$, in which case the wave minimal speed is called "pulled," and

Case 1(b): $c^*(f) > (2f'(0))^{1/2}$, in which case the wave of minimal speed is called "pushed."

Stokes [9], Rothe [8], Hagan [5] distinguish between these two sub-cases, and we shall see that our results do, also.

For the forced Fisher's equation

$$u_t = u_{xx} + f(u) + \mu g(x - ct), \tag{1.2}$$

with small μ and $g(x - ct)$ a travelling source, we will look for waves which are perturbations of a wave for (1.1).

Our results are:

Case 1(a): our methods do not apply, for all speeds c,

Case 1(b): near the pushed minimal wave U_{c^*} there is a simple bifurcation of nearby waves $U(c^* + \nu(\mu), z)$, but for nonminimal speeds $c > c^*(f)$ our methods do not apply, and

Case 2: when $g(z)$ is periodic, near the unique wave U_{c^*} there can be a transverse heteroclinic point in the (U, U') and thus "chaos."

In this short note we outline the results and the methods used to obtain them; details will be published in [10].

2. LINEARIZATION ANALYSIS

Wave solutions $u(x,t) = U(x - ct)$ of (1.1) satisfy

$$0 = U" + cU' + f(u), \quad ' = \frac{d}{dz} \tag{2.1}$$

and wave solutions $U(x - ct)$ of (1.2) satisfy

$$-\mu g(z) = U" + cU' + f(U). \tag{2.2}$$

Suppose that for $c = c_0$ there is a wave solution U_0 of (2.1). Then one can look for solutions of (2.2) of the form $U(z) = U_0(z + \alpha) + y(z)$, $c = c_0 + \nu$, so that $y(z)$ should satisfy

$$F(\alpha,\mu,\nu,z,y) = y" + c_0 y' + f'(U_0(z + \alpha))y, \tag{2.3}$$

where $F(\alpha,\mu,\nu,z,y) = -\mu g(z) - \nu y' + f'(U_0(z + \alpha))y - f(U_0(z + \alpha) + y) - f(U_0(z + \alpha))$. We use the method of Liapunov-Schmidt to analyze (2.3).

The linearization of (2.3) about $\mu = \nu = 0$, $y \equiv 0$, is

$$0 = y" + c_0 y' + f'(U_0(z + \alpha))y, \tag{2.4}$$

and one notes immediately that (2.4) has a solution $\phi(z) = U_0'(z + \alpha)$. Non-rigorous perturbation analysis of (2.3) leads one to the asymptotic

$$\nu \sim -\mu N^{-1} h(\alpha) \text{ as } \mu \to 0, \tag{2.5}$$

where

$$N = \int_{-\infty}^{\infty} e^{c_0 s} |U_0'(s)|^2 ds, \quad h(\alpha) = \int_{-\infty}^{\infty} e^{c_0 s} U_0'(s) g(s - \alpha) ds. \tag{2.6}$$

Unfortunately no sense can be made of (2.6) and thus of (2.5),

unless $U_0'(z)$ and $g(z)$ decay at appropriate rates as $z \to \pm\infty$.
When that decay is true the formal analysis makes sense, and
moreover, we can actually prove mathematical theorems about
that analysis. It is then that we say our methods do apply.

Define a norm

$$|y|_{\rho\omega} = \max\{\sup_{z\geq 0} e^{\rho z}|y(z)|, \sup_{z\leq 0} e^{-\omega z}|y(z)|\}$$

and Banach space $C_{\rho\omega} = \{y: \mathbb{R} \to \mathbb{R} \,|\, y,y' \text{ continuous and}$
$|y|_{\rho\omega} < \infty, |y'|_{\rho\omega} < \infty\}$. For some $\gamma > 0$, $\eta > 0$, the function
$\phi(z) = U_0'(z + \alpha)$ satisfies $\phi \in C_{\gamma\eta}$. In light of (2.6) it is
natural to expect that $2\gamma > c_0$ to be an important condition.

Lemma 1 [10, Lemma 3.1]: Assume (H0), U_0 solves (2.1)
for $c = c_0$, $G \in C_{\gamma\eta}$, and, most importantly, $2\gamma > c_0$. Choose
any γ',η' such that $\max\{0,c_0 - \gamma\} < \gamma' < \gamma$ and $0 < \eta' < \eta$.
Then

$$y'' + c_0 y' + f'(U_0(z + \alpha))y = G(z) \tag{2.7}$$

has a solution $y \in C_{\gamma'\eta'}$ if and only if

$$0 = PG \overset{\text{defn}}{=} N^{-1} \int_{-\infty}^{\infty} e^{c_0(s+\alpha)} \phi(s)G(s)\,ds. \tag{2.8}$$

If (2.8) holds, then one can define a linear continuous map
$K: C_{\gamma\eta} \to C_{\gamma'\eta'}: G \to y$.

We remark that in Case 1, where there are waves of all
speeds $c \geq c^*(f)$, the condition $2\gamma > c_0$ is satisfied only for
a pushed minimal wave.

Example: Hadeler and Rothe [4] found for $u_t = u_{xx} + u(1 - u)(1 + \sigma u)$, $\sigma > 0$ the explicit travelling wave
$U_0(x - c_0 t)$, where $U_0(z) = [1 + \exp((\sigma/2)^{1/2}z)]^{-1}$,

$c_0 = \gamma + \gamma^{-1}$, $\gamma = (\sigma/2)^{1/2}$. For $\gamma > 1$, i.e., for $\sigma > 2$, this wave is a pushed minimal wave.

In Case 2, Lemma 1 and more is true:

Lemma 2 [10, Lemma 3.3]: Assume (H0), (H2), and $G \in C_{00}$. Then there is a solution $y \in C_{00}$ of (2.7) if and only if (2.8) holds. If (2.8) holds, then one can define a linear continuous map $K: C_{00} \rightarrow C_{00}: G \rightarrow y$.

3. BIFURCATION

Theorem 3. Assume (H0), and assume either (H1) and $2\gamma > c_0$ or (H2). Then the method of Liapunov-Schmidt and the "bifurcation from a simple eigenvalue theorem" of Crandall and Rabinowitz imply there are solutions of (2.2) of the form $U(z) = U(\mu, \nu(\mu,\alpha), \alpha, z)$ near $U_0(z)$, where the function $\nu(\mu,\alpha)$ satisfies the asymptotic (2.5).

This theorem yields directly the result for Case 1(b) mentioned in Section 1. As for the result for Case 2 mentioned in Section 1, suppose $g(z)$ is periodic: Then one can use Theorem 3 along with the saddle point theory of periodic differential equations to obtain the existence of a transverse heteroclinic point, and thus infinitely many waves $U \sim U_0 + \mu \cdot$ (function periodic in z) and "chaos," if the Mel'nikov [1,7] function $h(\alpha)$ found in (2.6) satisfies some hypotheses.

REFERENCES

[1]   Chow, S. N. and Hale, J. K., Methods of Bifurcation
      Theory, Springer-Verlag, New YOrk, 1982.

[2]   Fife, P. C., Mathematical Aspects of Reacting and Dif-
      fusing Systems, Lecture Notes in Biomathematics #28,
      Springer-Verlag, Berlin, 1979.

[3]   Fisher, R. A., The advance of advantageous genes,
      Ann. of Eugenics 7 (1937), 355-369.

[4]   Hadeler, K. P. and Rothe, F., Travelling fronts in non-
      linear diffusion equations, J. Math. Biology 2 (1975),
      251-263.

[5]   Hagan, P., The Stability of Travelling Wave Solutions of
      Parabolic Equations, Ph.D. Thesis, Cal. Tech. 1979.

[6]   Kolmogorov, A. N., Petrovskii, I. G., and Piskunov, N. S.,
      A study of the equation of diffusion with increase in
      the quantity of matter, and its application to a bio-
      logical problem, Bjul. Moskovskovo Gas. Univ. #17
      (1937), 1-27.

[7]   Mel'nikov, V. K., On the stability of a center for time
      periodic solutions, Trans. Moscow Math. Soc. (Trudy) 12
      (1963), 3-56, MR 27-5981.

[8]   Rothe, F., Convergence to pushed fronts, Rocky Mt. J.
      Math. 11 (1981), 617-633.

[9]   Stokes, A. N., On two types of moving fronts in quasi-
      linear diffusion, Math. Biosciences 31 (1976), 307-315.

[10] Turyn, L., Travelling waves for forced Fisher's equation,
      Nonlinear Analysis (to appear).